FOOD FOOLISH

THE HIDDEN CONNECTION BETWEEN FOOD WASTE, HUNGER AND CLIMATE CHANGE

John M. Mandyck and Eric B. Schultz

A NOTE FROM THE AUTHORS

Hunger, food security, climate emissions and water shortages are anything but foolish topics. The way we systematically waste food in the face of these challenges, however, is one of humankind's unintended but most foolish practices. We wrote this book to call attention to the extraordinary social and environmental opportunities created by wasting less food. We are optimistic that real solutions to feeding the world and preserving its resources can be unlocked in the context of mitigating climate change.

A NOTE TO THE READER

Carrier, a unit of United Technologies, is a world leader in cold chain technologies for marine containers, road transport and supermarket display cases. On any given day, our technologies cool and preserve $6 billion in perishable foods on the open seas alone.

This long history of transporting, storing and preserving perishables gives us a unique perspective on the global food system. We don't just make "cold boxes" for our customers: We create affordable, environmentally advanced technology customized to a particular region and use. Our goal is to develop new, sustainable technologies that advance the global food supply chain to feed more people.

Over the last few years, as we focused intensely on our mission, we were surprised—and perhaps "stunned" is a better word—at the amount of food that is wasted globally. The impact of food waste on hunger, climate change, natural resources and food security is enormous. It's changing the way we think about our product and technology development. It's strengthening our commitment to sustainable innovation. It's also prompting us to convene research and food chain experts to find solutions.

We believe that food waste is an issue that must be elevated and examined globally.

That's why we published *Food Foolish*. It's not an attempt to be the final word on the topic of food waste. Rather, it's meant to connect the issues of hunger, resource conservation and climate mitigation. We hope it will be a catalyst for more meaningful global dialogue which, many think, is essential to the sustainability of the planet.

For updates related to the issues discussed in *Food Foolish*, please visit our website at foodfoolishbook.com. Proceeds from this book will be donated to food charities.

Together, we can secure the future of food.

TABLE OF CONTENTS

TABLE OF CONTENTS (CONTINUED)

FOREWORD

Many in the world take food for granted, while at the same time more than 800 million people go hungry every day. While avoiding food waste will go a long way toward fighting hunger, the environmental benefits are equally important. Consider water scarcity, widely considered one of the largest environmental challenges facing us in the 21st century. More water is used to grow food we throw away than is wasted in any other way by any country on Earth.

But waste is not confined to land, it is rampant in the ocean as well. Simply by fixing unsustainable fisheries we could feed up to an additional 1 billion people and save nearly $100 billion. Food is just one example of the myriad of resources and functions the oceans provide us, from oxygen to weather regulation. In many ways the oceans are a proxy for the problems we are trying to solve. They are a complex system and thus a powerful example of how we MUST connect the dots and recognize that all of these issues are related, from climate change, to energy, to fisheries to feeding the hungry.

Indeed, food waste would rank third in the world for greenhouse gas emissions if it was measured as a country. Yet, much like oceans, it is largely missing in the climate dialogue...that needs to change. Scientists estimate up to one-third of the carbon we emit is absorbed by oceans, causing them to become more acidic. This acidification of the ocean is having catastrophic effects on the food chain by limiting and in some cases preventing shell-based organisms such as plankton, oysters, crabs and coral from growing their shells. Today, in the Pacific Northwest alone, acidification is killing billions of sea creatures and costing hundreds of millions of dollars to local economies.

It is imperative that we raise awareness of the social, environmental and economic impact of food waste. It is a critical action that—as individuals, countries and the world community—we can take to improve the lives of people, combat climate change and help protect and restore our oceans.

–Philippe Cousteau

Cousteau is the son of Philippe Cousteau Sr. and grandson of Jacques Cousteau. He is an author, speaker, world traveler, social entrepreneur, founder of EarthEcho International and an Emmy-nominated television host.

FOREWORD

Food nourishes us with so much more than calories. It represents us and our cultures. It forms our traditions. It is our means of diplomacy. And as we gather together at the table in communion, we explore our relationships—not just with  each other, but with our planet. Together we can channel our collective will to find a sustained place in our world.

In my career, I've studied the confluence of food systems, human economies and health. While debate still nags for the definition of sustainability, I believe that regardless of specific language, our purpose is clear. Sustainability will be measured by the endurance of thriving human communities.

We currently suffer a great anxiety. We lose hope and doubt our ability to thrive: 7 billion of us in a world in which every day we ask ourselves, "Can I feed my neighbor?" My answer is yes. Yes, we absolutely can feed the needs of 7 billion people. But we also must recognize that we cannot feed the desires of 7 billion people. Need is finite and is written in our biology. Desire is infinite and is a product of our biography. We have incredible opportunity, in this decade, to begin to rewrite the biography of the human condition to support a sustainable relationship with nature, the nature that ultimately sustains us.

When food goes to waste, those who suffer first are the most vulnerable and the most invisible in our society. And while food insecurity is indeed a global issue, its impacts are all too often hidden. Food waste happens in every country. And in every country, every citizen bears civic responsibility to use our resources sustainably. It is important to acknowledge that we also have a responsibility to enjoy our resources.

When we waste food, we waste our opportunity to act as good neighbors. Food wasted is far more valuable than a simple measure of calories gone lost, or food that could be used to feed those who stand in the breadline. But that same food can be used to create opportunity and provide economic mobility for the very people who need it the most. Food and the opportunity it fuels can do more than feed the breadline—it can shorten it. Food not only feeds our bodies, it feeds the very soul of our society.

Our part is simple: smaller portions, served on smaller plates. Planning ahead to buy just what you need. Setting an example for peers by taking unfinished meals home from restaurants. It's not just a responsibility for us to purchase ugly, blemished foods, or to take home those leftovers. It provides us a much-needed opportunity to reflect upon all that we have and to reaffirm our will to find pleasure in it. We reflect not with guilt, but with hope. Because it is only with hope that together we celebrate the bounty in which we are all so fortunate to share on this beautiful planet.

—Barton Seaver

Seaver is a chef, Explorer with the National Geographic Society, and director of the Healthy and Sustainable Food program at the Center for Health and the Global Environment at the Harvard T.H. Chan School of Public Health.

INTRODUCTION: WHY WE SHOULD CARE

ONE-THIRD OR MORE OF THE FOOD WE PRODUCE EACH YEAR IS NEVER EATEN.

More than 1 billion metric tons of food is lost or wasted, never making it from the farm to our fork. Often in developing countries it decays in fields before harvest or spoils while being transported. Some is lost in retail markets before consumers can buy it. Meanwhile, in developed countries people buy too much and then throw it away. They reject perfectly nutritious food that is cosmetically imperfect. Too often we are served oversized meals, large portions of which are discarded. Although reasons vary, we waste food everywhere, often in ways that are unintended yet seem foolish given our fundamental need for this precious resource. In reality, we produce enough food to feed 10 billion people—everyone today and those expected by 2050. Yet people are still hungry.

It's hard to imagine so much waste of something so valuable in our modern, connected world.

OF THE FOOD WE PRODUCE
EACH YEAR IS NEVER EATEN

WHY SHOULD WE CARE?

More than 800 million people—a population equivalent to the United States and European Union combined—are chronically hungry. Two billion people, many of them children under 5 years old, suffer from malnutrition. Food waste[1] causes enormous misery and robs billions of people of their full potential.

MORE THAN

800 MILLION PEOPLE

A POPULATION EQUIVALENT TO THE UNITED STATES AND EUROPEAN UNION COMBINED— **ARE CHRONICALLY HUNGRY**

But the issues go well beyond hunger. In 2007 and 2008, an unexpected escalation in global food prices threw millions of people into poverty and caused panic in markets around the world. Riots rocked cities in 20 countries. The globe's urban populations—growing at 180,000 people *per day*—recognized just how far removed they had become from the sources of their food. Leaders around the world questioned their nation's food security.

The extraordinary growth of cities also means a rising middle class. This leads to demands for improved diets, shifting from simple grains to more nutritious meats, dairy, vegetables and fruits. The kinds of foods increasingly in demand are the very foods that reduce malnutrition. They are also the most susceptible to spoilage and waste and require the greatest care as they move through the food supply chain.

Food waste also has a devastating impact on the environment. The water used to grow just the food we discard is greater than the water used by any single nation in the world. Greenhouse gas (GHG) emissions are no less significant. The embodied carbon dioxide (CO_2) emissions in food waste alone represent 3.3 billion metric tons. That's all the energy that goes to produce the food we never eat, including fuel for tractors used for planting and harvest, electricity for water pumps in the field, the power for processing and packaging facilities, and more. In total, those emissions are more than twice the emissions of all cars and trucks in the United States. Viewed another way, if food waste were a country by itself, it would be the third largest emitter of greenhouse gases behind China and the United States. Yet the connection between food waste and

THE CONNECTION BETWEEN FOOD WASTE AND CLIMATE CHANGE IS MISSING

climate change is missing from policy discussions and public discourse.

Food conservation is every bit as important as energy conservation. Public policies have long and successfully encouraged energy efficiency to spread more power throughout our economy without having to build costly energy production facilities that result in greater environmental emissions. That same rigor must now be applied to prevent food waste.

The magnitude of food waste is shocking. Imagine purchasing three bags of groceries. While driving home, toss half of one bag of food onto the road. That represents the loss that occurs during harvest, processing and distribution. Arrive home and immediately toss the other half of the bag into the trash. That's the waste experienced by retailers and consumers. Buy three, get two: Welcome to our food system.

Some estimates show that we will need to increase global food production by 70 percent to meet the needs of our growing population. Yet we already produce enough food to feed everyone, including all those expected to join the planet by the middle of the 21st century. All we need to do is get that third bag of groceries home safely and onto our plates. When we waste less, we feed more. Even saving a portion of what is wasted can have a dramatic impact on reducing hunger, malnutrition, poverty, political instability, water shortages and carbon emissions.

There is a better way. Without action, the low-hanging fruit for reducing climate change will continue to literally rot before our eyes.

[1] Later we will distinguish between "food loss" and "food waste" as different problems with different potential solutions. For now, "food waste" is intended to describe all food mass lost along the food chain from farm to fork.

CHAPTER 1: THE ANATOMY OF HUNGER

WILL WE EVER RUN OUT OF FOOD?

In 1798, English economist Thomas Robert Malthus predicted that the world's population growth would eventually outrun its food supply. The results for humankind would be catastrophic. "The power of population is so superior to the power of the Earth to produce subsistence for man," Malthus wrote, "that premature death must in some shape or other visit the human race."[2]

For the next 150 years it seemed that Malthus just might be correct. In the 19th century some 25 million people died of famine, and more than 70 million perished in the 20th century.[3] In 1968, ecologist and demographer Paul Ehrlich became a modern-day Malthus in his best-selling book, *The Population Bomb,* when he re-emphasized Earth's finite capacity to sustain human civilization. "The battle to feed all of humanity is over," Ehrlich predicted. "In the 1970s the world will undergo famines—hundreds of millions of people are going to starve to death in spite of any crash programs embarked upon now."[4]

Fortunately, both Malthus and Ehrlich were wrong. Since *The Population Bomb* was published in 1968, the world's population has doubled to over 7 billion people. Despite this increase, humankind has managed to grow its food supply faster than its population. Eighty percent of the victims of famine in the last century died before 1965. Since the mid-20th century, famine has been more a function of civil disruption than of limited food supply.

"THE GREEN REVOLUTION"

What the doomsayers simply could not predict were advances in biology and genetics that would change the face of agriculture. In the 1940s biologist Dr. Norman E. Borlaug led research efforts on behalf of The Rockefeller Foundation to improve wheat production in Mexico. Over two decades Borlaug and his teams developed successive generations of wheat varieties with high yield potential. Similar work was done with rice. Combined with improved crop management techniques, new fertilizers and innovations in irrigation, these new, fast-growing, disease-resistant varieties transformed agricultural production in Mexico and later in Asia and Latin America. From the 1960s to the 1990s, for example, yields of wheat and rice in Asia doubled. The average Asian consumed a third more calories even as the population grew by 60 percent.[5]

This extraordinary increase in crop yield was coined by the U.S. Agency for International Development in 1968 as the "Green Revolution."[6] Dr. Borlaug was honored with a Nobel Prize and is described as the man who has "saved more lives than any other person who has ever lived."[7]

The Green Revolution is one of humankind's quiet miracles. From 1970 to 1990 the proportion of chronically underfed people in developing countries fell from 36 percent to 20 percent.[8] From 1975 to the late 1990s the percentage of underweight children under 5 years of age declined from 42 percent to 32 percent. "If we could compare all the 50-year periods in human history," the Food and Agriculture Organization of the United Nations (FAO) has concluded, "1950-2000 would almost certainly win first prize for speed, scale and spread of nutritional improvement."[9]

AFTER THE GREEN REVOLUTION

Even miracles have their limits, however. The Green Revolution relied on the ready development of affordable and fertile farmland. With much of the choicest land now being farmed, additional agricultural growth means a push into marginal lands and sometimes the destruction of important ecosystems like rainforests. The Green Revolution also relied on intense single-crop farming, which placed stress on land robbed of its biodiversity. By the 1990s crop yield improvement had fallen from 3 percent annually to 1 percent in developing counties.[10] Sub-Saharan Africa, which averages just one-sixth of the corn, wheat and rice yield of developed regions,[11] remained barely touched by the Green Revolution.

The specter of Malthus has risen again in the 21st century, well before humankind can guarantee an equitable and sustainable food supply. In a 2009 follow-up to *The Population Bomb,* Paul Ehrlich wrote that the most serious flaw in his book was that it was too optimistic about the future. Clearly it underestimated the impact of the Green Revolution, he wrote. And the absolute numbers of "hungry" people fell from their 1968 levels. But, he added, "the reduction of the hungry portion of the world population may well have been bought at a high price of environmental devastation to be paid by future generations."[12]

Punjab State, the breadbasket of India, is an example of challenges facing the agricultural community in the wake of the Green Revolution. Punjab is a region that more than doubled aggregate production of wheat and rice during the Green Revolution.[13] This success relied on hybrid grain seeds and intensive use of fertilizers and irrigation. Today, there is a shortage of suitable land for expansion. Plants are bumping up against their "photosynthetic limits" so that gains in yield are still possible, but not on the same order as during the Green Revolution. The stress on water, land and the environment placed by the need for more food is enormous as the population grows.

HUNGER TODAY

In the last 20 years the international community has come together to combat world hunger. The 1996 food summits in Copenhagen and Rome resulted in the shared goal of halving world poverty and undernutrition by 2015. In 2000, the *Millennium Declaration* recommitted the global community to the goal of halving the proportion of people who suffer from hunger.[14] In order to track progress, the FAO created the concept of Food Balance Sheets (FBS) to provide essential information on the food system of a country. Each FBS includes the domestic supply of food commodities, domestic food utilization, and the food supply available for human consumption.[15] From this country-by-country information, the FAO can estimate the prevalence of undernourishment at regional and global levels.

In short, the fantastic gains made in alleviating hunger and malnutrition in the last 50 years are slowing, and in some places reversing. Too many of the world's people remain hungry. And the dual pressures of a skyrocketing population and climatic changes threaten the future well-being of entire nations.

About 805 million people remain chronically undernourished.[16] Some 2 billion suffer from micronutrient deficiencies, also known as "hidden hunger." More than 100

805 MILLION PEOPLE REMAIN CHRONICALLY UNDERNOURISHED

MORE THAN

2 BILLION SUFFER FROM MICRONUTRIENT DEFICIENCIES (HIDDEN HUNGER)

100 MILLION CHILDREN UNDER THE AGE OF 5 ARE UNDERWEIGHT

million children under the age of 5 are underweight.[17] The vast majority of undernourished people live in developing countries where agricultural improvements have been uneven at best. Latin America and Southeastern Asia have experienced the greatest success in alleviating hunger. Western Asia has regressed since 1990. One in four residents of sub-Saharan Africa is chronically hungry, as are one-half billion people in Southern Asia.[18]

Malnutrition accompanies hunger. One quarter of all children under the age of 5 years globally—162 million—has inadequate height for their age, called "stunting," which can lead to diminished cognitive development.[19] In Mozambique, one of the world's poorest countries, almost half of all children under 5 are stunted by malnutrition.[20] Protein-energy malnutrition (PEM) is associated with the deaths of about 6 million children each year and is a universal cause of misery and death.[21]

The population grows while crop yields fall. Already suffering acutely, sub-Saharan Africa is the fastest-growing region in the world and expects population to increase from 926 million people to 2.2 billion by 2050.[22] On average, workforces in Asia and Africa are forecast to grow by 2 to 2.5 percent annually, while growth in crop yield has fallen to 1 percent.[23]

Developed countries are not immune from hunger. The FAO estimates that there are 14.6 million chronically hungry people in developed regions of the globe.[24] To differentiate from chronic hunger, the U.S. government began using the term "food insecure" in 2006 to describe households that lacked the money to ensure consistent access to adequate food. In 2013 that definition applied to 14.3 percent—*one in seven*—of American households.[25] Today, there are 50,000 emergency food programs around the United States to help ensure that those who are food insecure do not suffer from hunger.[26] Ironically, food insecurity can also lead to obesity in developed countries when people eat cheap food that is filling but does not provide balanced nutrition.

A NEW MIRACLE

In 2012, United Nations Secretary-General Ban Ki-moon issued a *Zero Hunger Challenge* to encourage a future in which people enjoy a fundamental right to food based on resilient and sustainable food systems.[27] The following year a summit of Latin America and Caribbean States leaders endorsed a hunger-free initiative for their nations. In July 2014 at the African Union summit, heads of state committed to end hunger on the continent by 2025. Countries like Bolivia have included a right to food in their constitutions.[28]

By 2050 we must feed more than 9 billion people. Many will enjoy a rising standard of living, which inevitably means a steady evolution from simple grains and starches to meat, dairy, fruits and vegetables. This will place an even greater strain on the global food system.

The political will exists to improve upon the gains of the Green Revolution, but the landscape has changed. While the focus remains on alleviating chronic hunger, there has emerged a fundamental understanding that simply expanding farmland and improving crop yields are insufficient to feed a growing planet. Any new solution must be sustainable. It must ensure a nation's food security so that each person has access to sufficient, safe and nutritious food. And it must employ the world's resources in ways that reduce agriculture's impact on climate change.

Observers agree that if humankind wants to engineer a new "miracle" to help feed our growing planet, it must be fundamentally different in shape and substance from the Green Revolution of the 20th century.

[2]Thomas Robert Malthus, *An Essay on the Principle of Population*, 1798, Chapter VII, p. 61.
[3]Stephen Devereux, "Famine in the Twentieth Century," IDS Working Paper 105, 2000, https://www.ids.ac.uk/files/dmfile/wp105.pdf, 9.
[4, 12]Paul R. Ehrlich and Anne H. Ehrlich, "The Population Bomb Revisited," *The Economic Journal of Sustainable Development*, 2009, http://www.populationmedia.org/wp-content/uploads/2009/07/Population-Bomb-Revisited-Paul-Ehrlich-2009b.pdf, 67.
[5]Tim Folger, "The Next Green Revolution," *The Future of Food (Special Compilation Issue)*, National Geographic Society, 2014, http://www.nationalgeographic.com/food-special-compilation/, 102.
[6]"Green Revolution: Curse or Blessing?", International Food Policy Research Institute, 2003, http://www.ifpri.org/sites/default/files/pubs/pubs/ib/ib11.pdf.
[7]Norman Borlaug Heritage Foundation, accessed 2015, http://www.normanborlaug.org/.
[8, 9, 10, 23, 27, 28]"The State of Food and Agriculture: Lessons from the Last 50 Years," Economic and Social Development Department, FAO Corporate Document Repository, 2000, http://www.fao.org/docrep/x4400e/x4400e11.htm.

[11, 20, 22]Joel K. Bourne Jr., "The Next Breadbasket," *National Geographic: The Future of Food (Special Compilation Issue),* National Geographic Society, 2014, http://www.nationalgeographic.com/food-special-compilation/, 34, 38, 37.

[13]Mark Doyle, "The limits of a Green Revolution," *BBC News,* March 29, 2007, http://news.bbc.co.uk/2/hi/6496585.stm.

[14, 15]FAO Food Security Statistics, accessed 2015, http://www.fao.org/economic/ess/ess-fs/en/.

[16, 18]FAO Hunger Map 2014, accessed 2015, http://www.fao.org/3/a-i4033e.pdf.

[17]G. Kennedy, G. Nantel and P. Shetty, "The Scourge of 'Hidden Hunger': Global Dimensions of Micronutrient Deficiencies," *Food Nutrition and Agriculture,* 2003, 8-16. Also, S.H. Wu et al, "Global Hunger: A Challenge to Agricultural Food, and Nutritional Sciences," Critical Review in Food Science and Nutrition, 2014, 54, 151-162. Also, "The State of Food Insecurity in the World 2012: Economic Growth is Necessary but Not Sufficient to Accelerate Reduction of Hunger and Malnutrition," FAO, Rome, 2012.

[19]"The Millennium Development Goals Report 2014," United Nations, http://mdgs.un.org/unsd/mdg/Resources/Static/Products/Progress2014/English2014.pdf, 4-5.

[21]"Food and Nutrition Security: Why Food Production Matters," 17.

[24]"The State of Food Insecurity in the World 2014," FAO, Rome, 2014, 40.

[25]USDA Economic Research Service, accessed 2015, http://www.ers.usda.gov/publications/err-economic-research-report/err173.aspx.

[26]Tracie McMillan, "The New Face of Hunger," *The Future of Food (Special Compilation Issue),* National Geographic Society, 2014, http://www.nationalgeographic.com/food-special-compilation/, 58.

CHAPTER 2: WHY ARE WE STILL HUNGRY?

IT'S DIFFICULT TO FIND A PLACE ANYWHERE IN THE WORLD WHERE SOMEONE ISN'T SUFFERING FROM SOME FORM OF HUNGER OR MALNUTRITION.

Houston, Texas, is the fourth largest city in the United States and has the fastest-growing economy of any U.S. metro area.[1] Its development is powered by energy, transportation, aeronautics and a strong manufacturing base. Only New York City is home to more Fortune 500 headquarters. Despite this unprecedented economic strength, a surprising number of Houston residents are hungry.

Brian Greene is president and CEO of the Houston Food Bank (HFB), a nonprofit organization that distributes food to nearly 600 hunger-relief programs in 18 southeast Texas counties. Through a combination of food pantries, soup kitchens and shelters, HFB helps feed 800,000 individuals each year.[2] "Hunger in America is complicated," Greene says. "We're almost never talking about starvation. We're talking about people missing meals. The phenomena we see most is that hunger is a way of saving money for another expense that you have to cover. People have to choose between food and rent," he adds, "or food and utilities. So hunger in America tends to be much more episodic."

The need is pressing for services provided by Greene's organization. In the Houston metropolitan area, 18.5 percent of residents are unable to consistently provide food for themselves and their families. Nearly one in four children is classified as food insecure. In fiscal year

2013-2014, HFB distributed 59 million nutritious meals and hopes to serve 100 million meals annually by 2018.[3] But it's not just about calories.

"When food banks started we were processed-food oriented," says Greene, looking back over his 30-year career. "We thought about cans and boxes—dry stuff. That's what was readily available. In the early days we were in the 'calories business.' If people were hungry and people needed calories, any calories were better than no calories." Since then, he notes, the food bank industry has matured and grown more sophisticated. "Most food banks don't see the world that way anymore," Greene adds. "The people that we're serving can have worse health problems than the general population, problems often associated with bad calories. If anything, we now need to be focused on good calories over bad calories. This fortunately correlates well with what is still probably the best opportunity for capturable waste in the U.S.," he concludes. "Fruits and vegetables. It's still billions that don't make it to market. We're trying to help capture that 'good calorie' waste."[4]

HFB is a member of Feeding America, the United States' largest domestic hunger-relief organization. Through a network of 200 food banks across the country, Feeding America nourishes 46.5 million people at risk of hunger, including 12 million children and 7 million seniors.[5]

America is hardly alone in its need. Hunger resides in affluent locations across the developed world. A welfare officer in Singapore reports that "Hunger is actually real here."[6] Willing Hearts, a welfare organization, began serving about 200 hot meals daily in 2005 and now serves 3,000 meals daily, seven days a week.[7] In London, England, the city's largest chain of food banks grew from just six in 2009 to 40 in 2013. Ninety-five percent of responding city teachers reported that children in their schools regularly went without breakfast, half of those because their family could not afford food. "Thousands of Londoners, both children and adults," a 2013 report concluded, "are in food poverty."[8]

THE CHALLENGE IN POOR AND DEVELOPING COUNTRIES

If feeding the hungry in affluent regions of the world requires careful planning and robust infrastructure, the complexity of the problem grows exponentially in developing regions.

As the world's single largest humanitarian organization fighting hunger worldwide, the United Nations World Food Programme (WFP) knows these challenges well. The WFP reaches about 80 million people every year.[9] The broad span of WFP programs reflects the tremendous complexity and stubbornness of hunger: a focus on women's economic opportunity, nutrition education, school meals programs, assistance with food procurement, improving logistics, and providing cash and vouchers to those who simply cannot afford to eat.

The WFP purchases more than 2 million metric tons of food every year for redistribution. Even when food is available, however, sometimes income is not. "Ethiopia was the first place in the world where WFP started distributing cash alongside food to refugees in camps," says

WFP's Country Director Abdou Dieng. "And we are seeing that even these modest sums of money are improving people's diets and self-esteem."[10]

Women on small farms account for 43 percent of the agricultural workforce in developing countries and play a vital role in improving agricultural productivity.[11] The WFP has found educating and empowering women to be among the most effective strategies to reduce hunger. "In order to really change a generation you need to make sure that every girl can read and write and count," says Catherine Bertini, former executive director of the WFP. "I don't think there is anything more important than that. For every year of school beyond a certain grade, a woman's income increases dramatically. For every year a sizable number of women are in secondary school, a country's GDP increases." She notes that school attendance also delays marriage. "Educated girls are less likely to carry AIDS. The longer they stay in school the less children they have. They're more productive in agriculture. They make more money. They're more likely to send their own children to school. Girls bring resources back to the family. There's only one initiative I would have to end hunger," Bertini concludes: "I would spend every bit of my money and political will to make sure that every girl was educated."[12]

Sometimes the way to reduce hunger in developing countries is through improved infrastructure, including roads and river transportation. Sometimes it is through the introduction of simple technologies. In 2009 in Afghanistan, a project of the Food and Agriculture Organization of the United Nations (FAO) provided 18,000 households with locally produced metal grain storage silos. These silos helped lower post-harvest insect and pathogen damage, reducing food loss from as much as 20 percent to 1-2 percent.[13] Other times simply improving the strength of storage containers can reduce crop damage while the use of low-cost evaporative and solar cooling technologies can preserve the shelf life of perishables.[14] "There are so many countries that produce a lot of something perishable when the price is low, and then a lot of it rots in the field or in the marketplace," Bertini explains. "When the season is over the price is high. Refrigeration systems and better storage systems—and a little more order in how things are grown—could be very helpful."

In some locations, reducing entrenched hunger requires private enterprise to create entirely new economic mechanisms. For example, Nigeria was the breadbasket of West Africa until oil reserves were discovered in 1958. Today, hunger in Nigeria is endemic. Nearly half of all children under 5 are undernourished.

Millions of farmers in Nigeria are "smallholders," growing crops on farms of just a few acres. Seed and fertilizer can be expensive and yields are one-fifth of those in more developed regions.[15] Corruption sometimes makes conducting business difficult. A third of all produce is wasted along dysfunctional supply chains.

A private farming program called Babban Gona is now helping Nigeria's subsistence farmers increase yields and improve their standards of living. In late 2011 Babban Gona began to "franchise" smallholders, providing training, loans, seed and fertilizer, tractors, and transportation at harvest. Babban Gona then warehoused and sold the grain, paying the farmers a dividend. Ibrahim Mustapha, one of Nigeria's smallholders, found himself for the first time with access to credit and affordable, high-quality seed and fertilizer. His maize yields improved to three times that of the average Nigerian farmer. This ensured his family was fed and put extra money in his pocket.[16]

Sometimes the solution in developing countries looks very much like that of developed countries. In 2010, the WFP began offering school lunches to children in Lesotho, a landlocked country where food insecurity affects a half-million people, or one quarter of the population.[17] Thirty-nine percent of children under 5 suffer from stunting. Four years later, the WFP provided school meals to nearly 250,000 children in schools throughout the country. An additional 190,000 children received two meals a day through a government-led program. The meals were designed to meet the micronutrient needs of youth, allowing them to learn and encouraging their families to keep them in school.

About 368 million children in 169 countries—one in five globally—receive a meal at school every day.[18] This makes school meals one of the largest hunger intervention strategies on the planet.

THE CHALLENGE IN RAPIDLY DEVELOPING COUNTRIES

Rapidly developing countries often have a growing middle class able to demand effective legislation and public services. Chile dramatically reduced its population of underweight children from 37 percent to 24 percent between 1960 and 2004 through the redesign of its health infrastructure. Political, academic and community organizations provided support. This included delivery of free milk, immunizations, and health and nutrition programs.[19] In Thailand, protein-energy malnutrition (PEM) was reduced from 36 percent to 13 percent between 1975 and 1990. A coordinated program included food supplementation, basic health care, investment in water and sanitation, and education.[20]

The nation of India operates one of the largest food safety nets in the world. Through the Targeted Public Distribution System (TPDS), the government aims to provide around 800 million people with subsidized monthly household rations each year. In 2013, India's National Food Security Act gave up to 50 percent of the urban and 75 percent of the rural population the legally enforceable right to state food benefits.[21] The TPDS has become an essential component of this fundamental guarantee.

In partnership with the World Food Programme, India's government re-envisioned TPDS as a best practices solution for subsidized food delivery. In 2014, WFP began work to assess the program's supply chain network and ultimately set up a door-to-door delivery system for food. Recommendations included simulation modeling, the adoption of a comprehensive supply chain management solution, electronic weighing and automated allocation.[22] The new TPDS model, which WFP has begun implementing in two states, would cost an estimated

$500-600 million if implemented across the country. The system is projected to save 8 to 10 percent of the current food subsidy.

HAVE WE DONE ENOUGH?

Ridding the world of hunger is a serious, complicated endeavor. Depending on the unique needs of a location, solutions can range from new roads, affordable seed and fertilizer, women's education, and nutritious school lunches, to food banks, transportation and storage technologies and cutting-edge software.

A pessimist might observe the Green Revolution slowing with more than 800 million people still hungry, a global population of 9 billion by 2050, and the impact of climate change growing. An optimist might see millions of lives saved and improved in the last half-century, new and innovative technology solutions, and government, not-for-profit and commercial entities galvanized around the issue of hunger as never before.

When U.N. Secretary-General Ban Ki-moon issued his *Zero Hunger Challenge* in 2012, it perfectly addressed these two distinct perspectives. "In a world of plenty, no one—not a single person—should go hungry," Mr. Ban said. His challenge included five main objectives which, if successfully met, would fix our broken food model. The first two were statements of principle: *Achieve 100 percent access to adequate food all year round,* and *End malnutrition in pregnancy and early childhood.* The third objective made clear that something different from the Green Revolution was now necessary: *Make all food systems sustainable.* The fourth cut a clear distinction between mass farming and farming by the masses: *Increase growth in the productivity and income of smallholders, particularly women.*

The last objective was something entirely new, something seldom even considered in 50 years of hunger mitigation: *Achieve a zero rate of food waste.*[23]

It is around this fifth, enormously compelling objective that our story continues.

[1]"America's Fastest Growing Cities 2015," Forbes, Jan. 27, 2015, http://www.forbes.com/sites/erincarlyle/2015/01/27/americas-fastest-growing-cities-2015/.

[2]Houston Food Bank fact sheet, accessed 2015, http://www.houstonfoodbank.org/media/103264/Houston%20Food%20BankBW%207-24-14.pdf.

[3]Houston Food Bank website, accessed 2015, http://www.houstonfoodbank.org/.

[4]Brian Greene, telephone interview with the authors, March 23, 2015.

[5]"Our History," Feeding America, accessed 2015, http://www.feedingamerica.org/our-response/about-us/our-history/?_ga=1.153946936.1240240951.1425245793.

[6]Maureen Koh, "Hungry in Singapore?", *AsiaOne News,* October 25, 2011, http://news.asiaone.com/News/AsiaOne+News/Singapore/Story/A1Story20111024-306837.html.

[7]"Welcome to Willing Hearts," accessed 2015, http://www.willinghearts.org.sg/welcome-to-willing-hearts/.

[8]"A Zero Hunger City: Tackling Food Poverty in London," London Assembly Health and Environment Committee, March 2013, http://www.london.gov.uk/mayor-assembly/london-assembly/publications/a-zero-hunger-city-tackling-food-poverty-in-london, 3, 15.

[9, 10]World Food Programme website, accessed 2015, http://www.wfp.org/.

[11]2014-2015 *Global Food Policy Report,* Washington, D.C.: International Food Policy Research Institute, 2015, 27.

[12]Catherine Bertini, telephone interview the authors, April 1, 2015.

[13]Tadele Tafera, "The Metal Silo: An Effective Grain Storage Technology for Reducing Post-Harvest Insect and Pathogen Losses in Maize While Improving Smallholder Farmers' Food Security in Developing Countries," International Maize and Wheat Improvement Center, accessed 2015, http://www.researchgate.net/publication/229293163_The_metal_silo_An_effective_grain_storage_technology_for_reducing_postharvest_insect_and_pathogen_losses_in_maize_while_improving_smallholder_farmers_food_security_in_developing_countries; also, Brian Lipinski et al, "Reducing Food Loss and Waste, *World Resources Institute,* working paper, June 2013, http://www.wri.org/publication/reducing-food-loss-and-waste, 101. This was part of a larger FAO program that distributed 45,000 metal silos in 16 countries which aided in storing 38,000 metric tons of grain. See "Food Losses and Waste in the Context of Sustainable Food Systems," a report by the High Level Panel of Experts on Food Security and Nutrition of the Committee on World Food Security, Rome, 2014, http://www.fao.org/3/a-i3901e.pdf, 59.

[14]Lisa Kitinoja, "Use of Cold Chains for Reducing Food Losses in Developing Countries," PEF White Paper No. 13-03, The Postharvest Education Foundation, December 2013.

[15]Francis Storrs, "The Solution to the Global Food Crisis Just Might Come From Nigeria," *Harvard Business School Alumni Bulletin,* March 2014, https://www.alumni.hbs.edu/stories/Pages/story-bulletin.aspx?num=3264.

[16]Babban Gona website, accessed 2015, http://www.babbangona.com/ibrahim-mustapha/.

[17]Tsitsi Matope, "International School Meals Day: When Two Meals Matter Most," World Food Programme, March 2015, http://www.wfp.org/stories/international-school-meals-day-when-two-meals-day-matter-most.

[18]World Food Programme website, accessed 2015, http://www.wfp.org/stories/school-meals-where-need-greatest-coverage-lowest.

[19, 20]"Global Framework for Action," World Food Programme and UNICEF, revised draft, December 2006, http://www.unicef.org/about/execboard/files/Global_Framework_for_Action1.0--Dec2006.pdf, 19.

[21, 22]"Targeted Public Distribution System: Best Practice Solution," World Food Programme, February 2014, http://documents.wfp.org/stellent/groups/public/documents/newsroom/wfp267097.pdf.

[23]"Rio+20: Secretary-General Challenges Nations to Achieve 'Zero Hunger'", UN New Centre, June 22, 2012, http://www.un.org/apps/news/story.asp?NewsID=42304#.VPsKovnF-3s.

CHAPTER 3: THE COMPETITION FOR LAND

NOBODY SAID OUR CHOICES WERE GOING TO BE EASY.

Cultivating land can mean food, health and security. Developing land creates industry, cities, commerce and a rising standard of living. Conserving land protects biodiversity, climate and the well-being of humankind. Yet doing all three simultaneously can generate enormous conflict. Never in the history of the world has the competition for land been more intense, nor the stakes higher. Some observers believe that the trade-offs we make among agriculture, development and conservation may be the most significant component of global change in the first half of the 21st century.[1]

THE COMPETITION FOR LAND

The Earth feels like a big place. Humankind has about 13.4 billion hectares of ice-free land surface, some 7.2 billion of which can sustain farming.[2] When it comes to increasing food production for a growing population, that seems encouraging. However, the details are more sobering.

AGRICULTURE—THE LARGEST HUMAN ENDEAVOR ON EARTH

USING **38.6**% OF OUR ICE-FREE SURFACE

Crop and livestock production already use about 5 billion hectares of land, 38.6 percent of our ice-free surface.[3] This is the richest, best watered and most easily accessible land on the planet. Of the remaining 2.2 billion hectares, only 1.4 billion is considered of "prime" or "good" potential.[4]

"Prime" and "good" are defined as any parcel of land that can support a single crop at just 40 percent of the maximum yield. However, this may not be the required crop. Thousands of hectares of land in North Africa are suitable only for olive trees—an important crop but not one alone that can feed a hungry world. The Earth's remaining "prime" and "good" farmland is also spread unevenly around the globe. Sub-Saharan Africa is blessed with 450 million hectares and Latin America 360 million. There is little additional quality farmland in the Near East, South Asia or Central America. In all, 13 countries account for 60 percent of the world's remaining viable agricultural land.[5] Some of this remains inaccessible due to civil unrest. The question of how much land we have to expand food production is one where a "big picture" perspective can be entirely misleading.

Of course, cultivating even the very best agricultural tract is filled with ecological peril. Some of Earth's potential farmland is fragile pasture likely to be converted to intensive, single-crop farming that endangers biodiversity. In sub-Saharan Africa between 1975 and 2000, land use for agriculture resulted in a 16 percent decrease in forests. During the same period, 83 percent of the agricultural expansion in the tropics came at the expense of forests.[6] Some prime potential farmland will also require considerable investment in infrastructure, including irrigation and connections to faraway markets.

Humankind has been farming for millennia. It should come as no surprise that we are left with those hectares that our ancestors avoided.

The race for the Earth's precious remaining farmland, including water and mineral resources, has become an outright global competition. This was aggravated by the 2007 and 2008 international food price spike, which created fear in a number of food-importing nations. Seeking food security, these nations acquired vast tracts of land, especially in Africa. Oxfam estimates that 227 million hectares of land changed hands in large-scale deals between 2000 and 2011.[7] Sometimes deals were done forcibly or skirted formal laws, giving rise to the term "land grabbing." This is a controversial global issue driven fundamentally by the fear of food insecurity.

WHEN CITIES COMPETE FOR LAND

The growth of cities is already one of the most spectacular stories of the 21st century. More land will be urbanized in the first three decades of the century than in all of human history.[8] Forecasters believe that by 2050, two-thirds of the planet's 9 billion people will live in urban areas.[9]

Today cities make up some 2.8 percent of the Earth's total land area.[10] The real footprint of a city, however, is measured by the intensity of its resource use, pollution, waste, climate impact and connecting infrastructure. It can also be measured by what it replaces: City sprawl inevitably triumphs over even the choicest of farmland.

China leads the world in new urbanization. The nation's city dwellers grew from 172 million in 1978 to more than 665 million in 2010. At just over half of its own population, however, China's urbanization rate is still below that of most developed countries. It's also far below its own national goal of 70 percent by 2050.[11] Meanwhile, the urban footprint in China more than tripled from 1981 to 2005 and is expected to triple again by 2030.[12] Since 2000, more than 70 percent of this increase replaced cultivated land. The government has taken recent measures to protect farmland, but urbanization has such a positive impact on standard of living and wealth creation that it is, in many ways, an unstoppable force.

ARE WE ALREADY IN DISTRESS?

The development of remote and satellite sensors has vastly improved our knowledge of Mother Earth. Studies are beginning to tell a troubling story of humankind's impact. One suggests nearly 24 percent of Earth's surface experienced a decline in ecosystem function and productivity between 1981 and 2003.[13] Another concludes that 40 percent of the land currently being farmed is already degraded.[14] The U.N. Millennium Ecosystem Assessment named land degradation among the world's greatest environmental challenges.[15]

Many responsible members of the global agricultural community have reacted by aggressively adopting sustainable farming practices. Even so, we may be asking too much. At our current rate of growth forecasters believe the Earth must support an increase of 60 to 120 percent in global crop demands by 2050.[16]

The silver lining is that these forecasts often make no allowance for our ability to curb the tremendous loss and waste that currently plague our food system. At a time when we are already pressing the limits of a distressed planet, preserving and consuming the food we currently produce offers a game-changing opportunity.

THE GLOBAL FOOD MARKET IS TRANSFORMING

As the global population grows to exceed 9 billion people, the kinds of food we produce and preserve become as important as the total amount of food produced.

Fruits and vegetables play an especially crucial role in global nutrition. The World Health Organization estimates 1.7 million deaths occur worldwide annually due to low consumption of these two food groups.[17] Such numbers are particularly tragic given that fruits and vegetables represent 40 percent of the 1.3 billion tons of food lost and wasted every year.[18] In

> **1.7 MILLION DEATHS OCCUR ANNUALLY DUE TO LOW CONSUMPTION OF FRUITS AND VEGETABLES**

part, this is due to the fact that one-fifth of the world's fruit and one-half of all vegetables are grown in Asia,[19] where the food delivery and storage systems are most fragile. It makes sense that we look first to fixing a broken food delivery system before we plant vast new tracts of land.

Cereals have always been at the heart of the world's food system. While their consumption peaked on a per capita basis in the mid-1990s, they still provide 53 percent of direct food consumption in developing countries and 49 percent around the world as a whole.[20] Their role as livestock feed is also growing.

In fact, the most pronounced dietary change in modern times has been termed the "Livestock Revolution." It occurs in increasingly wealthy populations that can afford to shift their diets from plant-based to animal products. The rapid rise of urbanization in the 21st century is just the kind of catalyst to drive this transformation. Forecasters now estimate that 40 percent of the world's population will participate in the Livestock Revolution by 2050.[21] This has critical implications in the competition for land.

Already, 75 percent of all crop and pasture land is dedicated to animal production.[22] Practiced responsibly, raising livestock is increasingly sustainable and meets the changing tastes of the global population. Practiced recklessly, it can lead to overgrazing, additional deforestation, and the creation of waste and GHGs. Increasingly, organizations representing food growers are taking the issue of sustainability seriously, addressing factors such as water quality, waste, climate impact, consumer health, animal welfare and even biodiversity.[23]

Along with beef, pork production and consumption is skyrocketing, especially in China. The country's total meat production, including chicken and beef, is expected to reach 93 million tons by 2020. Nearly two-thirds at retail weight will be pork. Since the mid-1970s, the average Chinese consumer's consumption of pork has grown by five times and is now nearly 50 percent greater than the average American's. In fact, there are about twice as many pigs in China as people in America.[24] Not surprisingly, pork consumption in this rapidly urbanizing country is a sign of wealth and well-being.

The impact of such growth is felt around the globe. More than half of the world's feed crops will soon be eaten by Chinese pigs. Brazil and Argentina have traded millions of hectares of forest to export soy and corn to China. Herbicide and antibiotic use has increased exponentially. Billions of tons of manure are creating environmental issues. One estimate shows that GHG emissions from all of Chinese agriculture increased 35 percent between 1994 and 2005.[25]

Production and consumption of poultry meat have also soared globally, growing from 15 percent to 32 percent of world meat production since 1970.[26] While more stressful on the environment than plant cultivation, poultry requires lower feed requirements per kilogram of meat than does beef.

As the world seeks to diversify sources of energy, crop production is also used for biofuels. Between 2000 and 2010, global biofuel production grew by 450 percent. The U.S. and Brazil combined to dedicate over 460 million metric tons of maize and sugar cane to biofuel production in 2010.[27]

In all, one study suggests that if the crop calories used for nonhuman uses were shifted to food for human consumption, existing agriculture could potentially feed about 4 billion more people.[28] In a complex world, simple shifts like that are not possible, or even desirable.

SURVIVING THE COMPETITION

The competition in agriculture between plant and animal production is the backdrop to the simple question of whether people around the world have enough good food to consume. One way of assessing progress is to track the number of calories consumed by a person on a daily basis (kcal/person/day). By that measure we have made good progress globally. Between 1970 and 2015, consumption improved from 2,373 kcal/person/day to an estimated 2,860.[29]

However, this masks some important regional trends. Latin America and the Caribbean have been essentially flat in the last decade. Developed countries consume 3,390 kcal/person/day while sub-Saharan Africa receives just 2,360 and South Asia 2,420. By 2050 both regions are forecast to remain 600 kcal/person/day below the 2015 consumption levels of developed nations.[30] That means that the fastest-growing regions in the world will continue to have the highest rate of undernourishment, lagging the developed world by more than a generation.

We must do better, and we can.

The demand for new land to meet the needs of agriculture, urbanization and conservation could be as high as 845 million hectares by 2030.[31] The global population is growing and our cities are exploding. Consumers are trading plants for livestock at mealtime. Biofuels are competing for lands used to feed people. With fruits and vegetables spoiling in fields and landfills, people are dying. Meanwhile, food demand through 2050 will increase three times faster than population growth.[32]

WASTE LESS: EASING THE COMPETITION FOR LAND

A study prepared in 2013 for the Millennium Institute[33] examined several scenarios by which agriculture production could adequately feed the global population in 2050.

A "business as usual" scenario forecast the need for a 45 percent increase in global crop production. This scenario showed food loss and waste growing 62 percent from 1.3 billion metric tons today to an unprecedented 2.1 billion metric tons in 2050. This additional 800 million metric tons of waste would add a staggering 2 billion metric tons[34] of carbon emissions to the environment annually.

A second scenario forecast a very modest reduction in food loss and waste from 2.1 billion metric tons to 1.85 billion metric tons in 2050.[35] This is still a startling number, but one that can unquestionably be reduced dramatically (as we'll discuss in later chapters). However, even under this highly conservative waste-reduction scenario, the benefits are still extraordinary.

First, the difference in saving 250 million metric tons of food is the equivalent of reducing carbon emissions by 600 million metric tons annually. Even more encouraging is an associated reduction of 132 million hectares of agricultural land required to feed the population in 2050.

That is the single most significant environmental payoff in reducing food waste: By moderating the growth of loss and waste through 2050, we eliminate the need to farm an area roughly the size of France, the U.K., Italy and Greece combined.[36] That means placing substantially less stress on land, water, energy, biodiversity and climate. That also reduces the pressure for desperate land grabs and improves global food security. Instead of food rotting, it also means more is sold by farmers, distributors and retailers.

Remember, too, that we may be able to reduce food

waste considerably below the numbers offered in even this improved scenario.

When it comes to food loss and waste, one important question is: How much better can we do? And for that we need to ask a second question: Where does all the good food go?

[1, 3, 10, 13]Roger LeB. Hooke et al, "Land Transformation by Humans: A Review," GSA Today, v. 22, no. 12, November 2012, http://www.geosociety.org/gsatoday/archive/22/12/pdf/i1052-5173-22-12-4.pdf, 4, 6.

[2, 4, 5, 20, 26, 29, 30]Nikos Alexandratos and Jelle Bruinsma, "World Agriculture Towards 2030/2050 (The 2012 Revision)," ESA Working Paper No. 12-03, Agricultural Development Economics Division, Food and Agriculture Organization of the United Nations, June 2012, http://www.fao.org/docrep/016/ap106e/ap106e.pdf, 10, 11, 42, 72, 23.

[6, 31]Eric F. Lambin and Patrick Meyfroidt, "Trends in Global Land-Use Competition," Rethinking Global Land Use in an Urban Era, Karen C. Seto and Anette Reenberg, eds., Strüngmann Forum Reports, Cambridge: MIT Press, Kindle edition, 2014, loc. 266, 253.

[7]The Global Land Grab: A Primer, TNI Agrarian Justice Programme, February 2013, http://www.tni.org/files/download/landgrabbingprimer-feb2013.pdf, 12.

[8, 32]Karen C. Seto and Anette Reenberg, "An Introduction," Rethinking Global Land Use in an Urban Era, Karen C. Seto and Anette Reenberg, eds., Strüngmann Forum Reports, Cambridge: MIT Press, Kindle edition, 2014, loc. 148.

[9]"World's Population Increasingly Urban With More Than Half Living in Urban Areas, United Nations Department of Economic and Social Affairs, July 10, 2014, http://www.un.org/en/development/desa/news/population/world-urbanization-prospects-2014.html.

[11, 12]Ziangzheng Deng et al, "Land-Use Competition Between Food Production and Urban Expansion in China," Rethinking Global Land Use in an Urban Era, Karen C. Seto and Anette Reenberg, eds., Strüngmann Forum Reports, Cambridge: MIT Press, Kindle edition, 2014, loc. 1006-1019, 1033.

[14, 15]Ian Sample, "Global Food Crisis Looms as Climate Change and Population Growth Strip Fertile Land," The Guardian, 2007, http://www.theguardian.com/environment/2007/aug/31/climatechange.food.

[16, 21, 22, 27, 28]Emily S. Cassidy et al, "Redefining Agricultural Yields: From Tonnes to People Nourished Per Hectare," Institute on the Environment, University of Minnesota, IOP Publishing Ltd., 2013, http://iopscience.iop.org/1748-9326/8/3/034015/pdf/1748-9326_8_3_034015.pdf, 1, 2, 4.

[17]"Promoting Fruit and Vegetable Consumption Around the World," World Health Organization, 2015, http://www.who.int/dietphysicalactivity/fruit/en/index2.html.

[18]Food Wastage Footprint: Impacts of Natural Resources, Technical Report (Working Document), FAO, http://www.fao.org/3/a-ar429e.pdf, 103.

[19]"FAO Statistical Yearbook 20013: World Food and Agriculture," Rome, 2013, http://www.fao.org/docrep/018/i3107e/i3107e00.htm.

[23]Tyler Harris, "How Sustainable Is Livestock Production," Farm Futures, February 7, 2013, http://farm-futures.com/story-how-sustainable-livestock-production-18-94448.

[24]Sam Brasch, "How China Became the World's Largest Pork Producer," *Modern Farmer,* March 11, 2014, http://modernfarmer.com/2014/03/tail-curling-facts-chinese-pork/.

[25]"Empire of the Pig," *The Economist,* December 20, 2014, http://www.economist.com/news/christmas-specials/21636507-chinas-insatiable-appetite-pork-symbol-countrys-rise-it-also.

[33]"Global Food and Nutrition Scenarios: Final Report," Millennium Institute, Washington, D.C., March 15, 2013, http://www.un.org/en/development/desa/policy/wess/wess_bg_papers/bp_wess2013_millennium_inst.pdf.

[34]The current relationship determined by the FAO is that 1.3 billion metric tons of lost and wasted food results in 3.3 billion metric tons of CO_2 equivalent. That relationship is applied here: 800 million x 3.3/1.3, or 2 billion.

[35]The report's scenario for reduced loss and waste does not appear to address possible improvements to the harvest and post-harvest stages of the food supply chain. The focus is limited to processing, distribution and consumption. Substantial gains around food loss can be made in these early stages of the food supply chain.

[36]"List of Countries and Dependencies by Area," Wikipedia, http://en.wikipedia.org/wiki/List_of_countries_and_dependencies_by_area. France's land area is 640.4 thousand square kilometers, Italy's 294.1, the UK's 241.9 and Greece's 130.6, or 1.31 million square kilometers.

CHAPTER 4: ENTER THE COLD CHAIN

CONSIDER THE MAGNIFICENT BANANA.

It is humankind's favorite fruit and one of the most important crops in the world. About $44 billion worth of bananas are grown in 130 countries, employing hundreds of thousands of people.[1] Bananas contain vitamins B-6 and C, manganese, potassium and other minerals essential to human health. They provide 400 million people in some of the world's poorest places with up to one quarter of their daily calories.[2]

Consumers in developed nations also love bananas. The European Union is the largest importer in the world.[3] Americans buy more bananas than apples and oranges combined,[4] and eat 27 pounds of the yellow fruit per person each year.[5]

The problem with bananas, of course, is that they are highly perishable. Too hot and they spoil. Too cold and they blacken. Mishandled and they bruise. Adored by insects, pests and pathogens of all kinds, bananas ripen from the moment they're picked to the moment they're consumed. And there are lots of bananas on the move every year; some $8.9 billion[6] worth of exports make them the fifth most-traded agricultural commodity behind only cereals, sugar, coffee and cocoa.[7]

THE TRAVELING BANANA

Needless to say, bananas in transit require one of the most sophisticated and impressive delivery systems in the modern world. It's called the cold chain.

The life of a traveling banana looks something like this: Hand-picked in large bunches when they are green, the fruit is padded and wrapped for protection, sometimes hauled from the plantation by mule, and delivered as quickly as possible to a packing station. There each bunch is washed, inspected, cut, treated, and carefully boxed for transport. The journey from field to box might take two or three days, all while the banana is ripening.[8]

From there the boxes are placed on trucks or in shipping containers and delivered to container ships. Transported by sea under precise conditions and arriving at the destination port, they are unloaded, shipped to cold distribution centers (and sometimes to special banana ripening rooms), unpacked and inspected, and reshipped in refrigerated trucks and trailers to a warehouse, and then again to a retailer. Inspected again, they are finally placed on shelves for sale.

Sound complicated? From hand to packing plant to container to ship to truck to warehouse back to truck to retailer to consumer—there are frequent careful handoffs. Five or six different entities might be involved. Information and money flow back and forth. The entire journey can take two to three weeks.

When all goes well, the bananas are lovingly maintained throughout their travels at an optimal temperature. Shoppers in developed countries find their fruit at the end of this long journey perfect and ready to eat.

THE MODERN COLD CHAIN

Perishable foods can spoil quickly, and it's often not

a pretty sight. For instance, the flesh of a healthy fish is sterile but also a perfect medium for bacteria. Some bacteria may come from the fish's skin or scales while being prepared. Other bacteria can be transferred from sea mud or the fish's intestines. Whatever the source, bacteria can multiply rapidly in hot conditions, turning pristine, edible flesh to a mushy, sour-smelling, inedible mess. In fact, fresh fish at 30°C/86°F spoils in just a few hours. Likewise, as vegetables begin to decompose, they can become active breeding grounds for bacteria, mold, yeasts and other harmful microorganisms. Fresh green vegetables will last less than two days at 30°C/86°F.

FRESH GREEN VEGETABLES WILL LAST LESS THAN TWO DAYS AT 30°C/86°F

There are many ways to preserve a perishable food. Our ancestors tried them all, from drying to salting and canning. However, nothing keeps perishable products safe, maintains their physical and nutritional qualities, and prolongs their shelf life like cold air. Fresh fish held at optimal temperature can last for 10 days; fresh green vegetables might last a month at optimal temperatures.[9] Drive one or the other to market in an open truck in the hot sunshine and they can be ruined before arrival.

That makes the modern cold chain an indispensable tool of global trade and increasingly a field of high technology. This seamless and interconnected network includes marine container refrigeration, truck and trailer refrigeration, warehouse and food retail refrigeration, and millions of home refrigerators. It involves information systems, trained experts, and superb cooperation and communications. The cold chain quietly protects the food supply and raises the standard of living. In much of the developed world it has been so effective for so long that its magic is taken for granted. More than just adding to consumer variety and supporting agricultural jobs, the cold chain plays a major role in ensuring food safety, often meeting rigorous standards to the benefit of customers. The Institute of Mechanical Engineers has determined that transport refrigeration *alone*—just one segment of the modern cold chain—could avoid a quarter of food waste in developing countries.[10]

In 1900 the cold chain was virtually nonexistent. A century later, some 550,000 marine containers transported food throughout the world to approximately 1.2 million refrigerated road vehicles.[11] Today, the modern cold chain continues to expand, protecting everything from meat and dairy to fruits and vegetables. In parts of the world where the cold chain is established and effectively run, perishable food loss can be maintained as low as 2 percent.[12]

IN PARTS OF THE WORLD WHERE THE COLD CHAIN IS ESTABLISHED PERISHABLE FOOD LOSS CAN BE MAINTAINED AS LOW AS 2%

Our tale of the traveling banana also highlights the fact that the cold chain preserves not just food, but especially nutritious food. As consumers grow wealthy we know their consumption of starchy foods declines[13] in favor of produce, meat, fish and dairy. These highly perishable foods are absolutely dependent on an effective cold chain. A report by the International Institute of Refrigeration concluded that "Greater use of refrigeration technologies would ensure better worldwide nutrition, in terms of both quantity and quality."[14]

Consequently, the modern cold chain is not just a collection of cooling technologies that preserve perishable products, but an intensive, value-added process that seeks to extend shelf life, protect the safety and integrity of products, reduce food loss and enhance global food security. While offering these benefits, the cold chain is becoming more energy-efficient by utilizing more environmentally sustainable technologies, including natural refrigerants to lower greenhouse gas emissions.

Such a powerful tool should be universally deployed. Unfortunately, only about 10 percent of perishable foods today are refrigerated worldwide.[15]

THE IMPACT OF THE COLD CHAIN

Consider again our banana.

India grows 28 percent of the world's crop and is the top-producing nation in the world.[16] However, like many rapidly developing countries, India does not yet have an extensive or mature cold chain. Of the 104 million metric tons of perishables transported

INDIA:
THE WORLD'S BANANAS SUPPLY

PRODUCES **28%**

EXPORTS **0.3%**

An upgraded cold chain infrastructure could:

INCREASE
3,000 ▶ 190,000
CONTAINERS EXPORTED

PROVIDE
95,000
JOBS

&

BENEFIT
34,600
SMALLHOLDER FARMERS

among cities throughout India, only 4 million are transported in refrigerated vehicles, and that's mostly milk.[17] India's 6,000 cold storage units accommodate only about 11 percent of the country's crops, most of this potatoes.[18]

Without temperature control, observers believe that India loses 20 to 50 percent of its total production of fruits and vegetables each year.[19] That makes local delivery of a highly perishable fruit like bananas wasteful, and export simply impossible. Of India's 28 percent share of the world's production, the country exports less than 1 percent.[20] Observers believe that with an advanced cold chain, India could grow its banana exports from 3,000 to 190,000 containers. This would benefit 34,600 smallholder farmers.[21] Reducing losses of this especially nutritious food could alternatively reduce hidden hunger and ensure food security for the country's growing population.

The reach and impact of the modern cold chain are extraordinary but terribly uneven. In the United States, 70 percent of all the food consumed each year passes through this seamless refrigerated network.[22] In a rapidly developing nation like China, equivalent cold chain coverage is 25 percent for beef and 5 percent for produce.[23] In parts of the world like sub-Saharan Africa, the installation of basic cold chain practices is hindered by poor roads and infrastructure. Some 2.6 billion people worldwide who lack affordable and reliable energy services are simply beyond a functioning cold chain.[24]

Rapid urbanization represents an

AS THE WORLD'S POPULATION MIGRATES TO CITIES, THE DISTANCE BETWEEN FOOD PRODUCTION AND CONSUMPTION GROWS

additional complicating factor. As the world's population migrates to cities, the distance between food production and consumption grows. Dependence on an effective cold chain becomes ever greater.

India is investing aggressively in refrigerated warehouses and transport. One example is Coldex Logistics, a leading cold storage transport company based in the historic city of Gwalior, some 200 miles south of Delhi. Coldex was founded as a dry storage trucking company and has evolved over the last decade to support the cold-chain needs of a variety of international restaurant brands and confectionery companies operating in India.[25] "We invested in the cold chain to help our customers reach out to their customers," says Managing Director Gaurav Jain. "While doing so we have enabled the larger aspect of controlling wastage by prolonging the shelf life of raw and processed food." Coldex is now expanding beyond transport to warehouses, adopting increasingly sophisticated IT systems, all with an eye toward managing every link in the cold chain. "And very clearly that's the path a developing nation like India is taking to eradicate food waste and thus hunger and poverty," adds Mr. Jain. "We feel this can only be done by efficient and functional integrated supply chains—from farm to fork."[26]

Likewise, China is addressing its need for a modern cold chain. The Chinese grocery market exceeds $1 trillion, the largest on Earth. Yet most fruits and vegetables grown in the country are transported in open trucks, resulting in losses that can exceed 40 percent.[27] As China adjusts to meet the needs of its growing urban middle class, the country's cold chain logistics industry is expected to grow at 25 percent annually. The first set of investments is being made in cold warehouses. Once there is a place to store perishables, China will invest in refrigerated trucks.[28]

In Latin America, supermarkets have grown segment share from as low as 10 percent to a current 60 percent by making massive investments in the cold chain.[29]

Globally, the total capacity of refrigerated warehouses grew 20 percent to 552 million cubic meters between 2012 and 2014.[30] India is the world leader with 131 million cubic meters, followed by the U.S. with 115 and China with 76. Compared with the population, however, the U.S. has 360 cubic meters of refrigerated storage capacity for every 1,000 inhabitants while India has 102 and China 54.[31] Were China and India to expand cold chain intensity to U.S. levels, global refrigerated warehouses would grow from 552 million cubic feet to 1.3 billion.

WHEN THE COLD CHAIN WORKS

Turek Farm is the third largest vegetable farm in the state of New York and a top 10 grower nationally of sweet corn. David, Frank Jr. and Jason Turek also grow squash, pumpkins, cabbage and green beans on 4,000 acres of rain-fed farmland. Everything grown is perishable and each product will make its way successfully to market up and down the U.S. East Coast if, and only if, the company's carefully managed cold chain does its part. "A lot of people can make a decent crop," says Jason Turek. "Getting it to market safely can be a challenge."[32]

Of course, Jason is being modest. Farmers like

the Tureks face everything from national competition, inferior seed and pests to drought and flood. "Even the weather has become more extreme," adds Jason, feeling the effects of climate change. "Normal is now the average between two extremes." Waste can begin at the moment a seed is planted; if it happens to be a half-inch deeper than those around it, the vegetable might ripen too late to be harvested with the rest of the crop. With so many obstacles to profitable farming, so much investment at stake

> **"NORMAL WEATHER IS NOW THE AVERAGE BETWEEN TWO EXTREMES"**

each season, and increasing demands from retailers, the need to preserve healthy produce once harvested is paramount. This requires a robust, unbroken cold chain.

"Sweet corn quality can be lost overnight without refrigeration," says Frank Turek Jr., and "it'll taste like cardboard."[33] Consequently, once picked, corn on the Tureks' farm is quickly packed and cooled to about 4.4°C/40°F. From there it's moved to refrigerated storage, onto refrigerated trucks (with a covering of ice on each truckload to preserve moisture), and eventually to a refrigerated retail distribution center. If all goes well, an ear of Turek sweet corn won't rise much above 4.4°C/40°F until it reaches the shelves of a grocery store.

Having a seamless cold chain is the only way the Tureks can preserve freshness and shelf life of their products, ensure food safety, and meet the specifications of retailers. It is the only way their product can delight customers. And perhaps most importantly, their investment in a modern cold chain is the only way to reduce food waste, a specter that haunts farmers—even those with access to the best agricultural inputs and equipment available—from the moment they plant seed in the ground. Despite decades of success and a strong brand, Jason says, "We're only as good as our last order."

THE BROADER FOOD SUPPLY CHAIN

Not all foods need to be handled quite as carefully as vegetables, fruits, dairy and meat. But all food is perishable and subject to loss.

Take the example of grains, which provide the world with over one-half its calories and are an important source of feed for livestock. In 2012 the World Bank and FAO studied the "grain chain" of wheat imports to Arab countries.[34] This subtropical region is the largest net importer of cereal calories in the world, buying about 56 percent of what they consume.[35] When cereal prices are volatile or supply is at risk, food security in the Arab world can become an issue of national security.

Like fruits and vegetables, an efficient supply chain for wheat can lower cost of the product to consumers and reduce waste. The World Bank study found that wheat losses in Arab countries were as much as 5 percent, leading some nations to import more than they needed.[36] Inefficiencies varied by country but included long ship turnaround times in port, idle trucks and mills, long transit times due to poor roads, poor handling systems, spillage, spoilage, pilferage, and limited or outdated storage capacity. The average time from unloading at the port to bulk storage at the flour mill in 2009 was 78 days in Arab countries compared with 47 days in South Korea and 18 days in the Netherlands.[37] The study suggested a number of possible improvements. It also highlighted that even when a food product like cereal does not have the perishability of bananas, it still travels a complicated supply chain that requires investment and expertise to function optimally.

A similar study of the supply chain for grain in sub-Saharan Africa found losses from farm to processing could range from 10 to 20 percent. Some losses resulted from harvesting methods, handling procedures, types of storage, pests or pathogens. Sometimes the supply chain was impeded by mismanagement or political problems.[38] In total, such losses are devastating. Research suggests that post-harvest losses in sub-Saharan Africa over the last decade exceeded the total food aid received by that region and could have fed an additional 48 million people. Given these findings, some experts now believe that investments in reducing post-harvest losses are "quick impact inter-ventions" for enhancing food security.[39]

The truth is, every food supply chain is simply some variation of the cold chain. All perishables need to be shielded from temperature and moisture extremes. The sooner they are consumed after harvest or slaughter, the safer, more appetizing and more nutritious they tend to be. Companies and countries that have become skilled at moving meats, dairy, fruits and vegetables—protect-ing their "traveling bananas"—are well on their way to reducing needless loss and enhancing their total food security.

THE SUSTAINABLE COLD CHAIN

Even great benefits come with some cost, and expanding the cold chain is not without its trade-offs. The refrigerated trucks, trailers, shipping containers, warehouses and retail displays that comprise the cold chain all require energy and are a source of global hydrofluorocarbon (HFC) refrigerant emissions. Fortunately, as the cold chain evolves and expands, so too do the technologies available to reduce its environmental footprint. In particular, advances are being made in non-ozone depleting, low global warming and natural refrigerants, fuel reduction technologies for road transportation systems, and improved recyclable materials.

Kevin Fay is executive director of the Alliance for Responsible Atmospheric Policy, an organization focused on industry participation in the development of International and U.S. government policies regarding ozone protection and climate change. Under the Alliance, member companies and others have committed $5 billion over the next decade to research, develop, and commercialize low global warming potential technologies.[40]

Fay is also the executive director of the new Global Food Cold Chain Council. "We're making great progress in greening the cold chain," Fay says. "We're using a broader base of refrigerants and placing a greater reliance on things like information technology that are focused on helping cold chain equipment perform more efficiently. There has been a tremendous advance in the development and efficiency of cold chain equipment."[41]

The Global Food Cold Chain Council is a promising example of how the private sector can play a leading role in innovating climate-sustainable technologies that advance a sustainable cold chain. Announced by a coalition of major companies at the special United Nations Climate Summit 2014, the council works with partners in the Climate and Clean Air Coalition to develop and implement broad-based public and private sector collaborative solutions to reduce HFC emissions in the cold chain across developed and developing countries.

A sustainable cold chain and its impact on food waste is now poised to change the climate conversation. "The initial focus on greening the cold chain started with improving refrigerants," Fay says. "That led to consideration of the energy efficiency issues around the entire cold chain, which are much more important over the life of the equipment. Subsequently, this led us to the issue of food waste. The cold chain has the ability to significantly reduce food waste in processing, transport and retail," Fay adds. "So our efforts started as a focus on refrigerant usage but have now broadened to energy efficiency and beyond that to significantly reduce food waste by additional reliance on cold chain technology."

Retail refrigeration is one link in the cold chain making significant moves away from traditional HFC direct-expansion systems to embrace natural or low global warming refrigerant technologies. Retailers are leading the way. CO_2 as a refrigerant for supermarkets and convenience stores is economical, environmentally sustainable, safe and energy-efficient. Legislation is helping drive adoption, with countries like Denmark taxing HFC

CO₂ FOR REFRIGERATION IS ECONOMICAL, ENVIRONMENTALLY SUSTAINABLE, SAFE AND ENERGY-EFFICIENT

use and Germany providing direct financial incentives for using low global warming refrigerants. In other cases, consumer pressure and environmental groups have driven adoption. Research indicates that among European food retailers, 69 percent have an enterprise or group-level strategy to encourage carbon footprint reduction, and more than half would reduce normal investment cycles to promote the use of more environmentally balanced refrigeration technology.

In the U.S., the public and private sectors joined forces in September 2014 around a series of commitments designed to reduce the cumulative consumption of greenhouse gases by the equivalent of 700 million metric tons of CO_2 by 2025—the same as taking nearly 15 million cars off the road for 10 years.

Fay cites the willingness and ability of the world's large developing countries to embrace modern cold chain technology as critical to the growth of a sustainable cold chain. "On the equipment side, we've got traction throughout the developed country economies," he says. "That's rapidly spreading to the large developing country economies such as India, China and Brazil. This has been our initial focus, on the equipment side. There's been a great deal of history and a great deal of success in dealing with technology transfers that are both environmentally protective and economically sensible. Our efforts," Fay adds, "are to build on that successful process. What's happening now is the transition from just an equipment focus to the much broader issues associated with food waste."

There are challenges. "The attitude in some countries is to just produce more food," Fay says. "Transportation issues can be difficult. Retail systems are not advanced. And so the first reaction is not to worry about the waste—it's just food waste. What could be harmful about that?" Fay muses at his own rhetorical question. "So there's a huge educational process for policymakers, people in the agricultural community, and people in the food chain to understand that relying on the cold chain is going to be much more productive, much more beneficial to the population, and much less impactful on other resources that are also stressed, such as water supply."

"There's a great deal of education required," Fay concludes, "in order to be effective. It's showing them how they can do something different, something that's cost-effective, and something that has both societal and environmental benefit."

As retail refrigeration improves, so too do the transport links. The Pacific Northwest National Lab (PNNL) in Richland, Washington, a Department of Energy research lab, is working to power truck refrigeration units with hydrogen. Other companies are adopting electric delivery trucks that use electric chillers instead of conventional diesel-powered units. New initiatives include installing energy-efficient production lines, new coolers with energy management devices, and reduced transportation through improved backhauling.[42] Several large cold chain players have teamed up to form "Refrigerants, Naturally!" to promote a shift in point-of-sale cooling technology toward natural refrigerants with a low- or non-Global Warming

Potential and a zero Ozone Depletion Potential.[43]

One report suggests that refrigeration is an area where "dramatic emission cuts could be made relatively easily" simply by using and maintaining energy-efficient equipment correctly.[44] In other words, just as we saw that awareness and education could have a substantial impact in reducing consumer food waste, proper training can play a major role in improving commercial performance.

New technologies will be an important element of the sustainable cold chain. CO_2 is especially promising. Refrigeration units are now available for marine container application that run on CO_2 and can reduce carbon footprint by 28 percent compared with previous equipment. CO_2 for supermarket refrigeration is currently installed in more than 1,500 supermarkets across Europe. The cold chain is also harnessing the power of the sun to charge transport refrigeration unit batteries to maintain peak performance in an environmentally sustainable way. Installed on the roofs of trailers, truck bodies

NEW TECHNOLOGIES WILL BE AN IMPORTANT ELEMENT OF THE SUSTAINABLE COLD CHAIN

and refrigerated rail cars, and exposed to daylight, these flexible solar panels ensure power for system starts and avoid costs associated with weak or dead batteries. They also conserve fuel by minimizing the need to run the refrigeration unit engine to charge the battery.

Biodegradable and recyclable packaging materials are increasingly being looked at to provide more environmentally balanced solutions that do not compromise the safety of products. Reducing the overall size and weight of the packaging is also seen as a way of making cold chains greener, while reducing transport costs.

Of course, an expanded cold chain is only one of several strategies to reduce food waste. More research is needed to understand the net environmental benefit of an expanded cold chain and its ability to reduce the greenhouse gas emissions associated with food waste detailed in chapter 6. In addition to new, sustainable refrigeration technologies, the environmental footprint of the cold chain will continue to benefit from a world that moves to cleaner sources of energy.

The greening of the global cold chain "is not a short-term play," Fay concludes. "We've been in the climate policy debate globally for more than two decades now. This food waste and cold chain discussion will go on for decades in terms of climate policy as we continue to expand the equipment opportunities, the environmental achievements, and the benefits that society will gain."

A COLD CHAIN REVOLUTION?

The cold chain is a phenomenon of the modern world, but only a fraction today of what it might someday become.

Where the modern cold chain exists, it can be substantially improved. Where it is still nascent, it can be expanded. Where it does not exist, it can be built. Education and technology continue to make it more sustainable. Around the globe, investment in an integrated network of temperature-controlled space will have a profound impact on feeding and healing a hungry world and helping farmers get more product safely to market. All without planting a single new hectare of land.

The world has witnessed a Green Revolution. It is now experiencing a Livestock Revolution. Perhaps what humankind needs next is a "Green" Cold Chain Revolution.

[1, 16]Edward Evans and Fredy Ballen, "Banana Market," Food and Resource Economics Department, UF/IFAS Extension, February 2012 (reviewed January 2015), http://edis.ifas.ufl.edu/pdffiles/FE/FE90100.pdf.

[2, 4, 6]Gwynn Guilford, "How the Global Banana Industry is Killing the World's Favorite Fruit," Quartz, March 3, 2014, http://qz.com/164029/tropical-race-4-global-banana-industry-is-killing-the-worlds-favorite-fruit/.

[3]"Europe Largest Banana Importer Worldwide," *Fresh Plaza,* April 25, 2013, http://www.freshplaza.com/article/108375/Europe-largest-banana-importer-worldwide.

[5]Marisa Taylor, "Chiquita Merger Reignites Fears of a Disappearing Banana Crop," Al Jazeera America, March 10, 2014, http://america.aljazeera.com/articles/2014/3/10/chiquita-merger-createsworldsbiggestbananacompany.html.

[7]"Bananas," Fairtrade International, 2011, http://www.fairtrade.net/bananas.html.

[8]"Dole-Harvest Bananas," YouTube, April 3, 2013.

[9, 15, 17, 18, 19, 20, 21, 24]C.G. Winkworth-Smith et al, "The Potential Value of Reducing Global Food Loss," The University of Nottingham, Division of Food Sciences, School of Biosciences, March 2015, 17, 19.

[10]http://www.theguardian.com/sustainable-business/2014/dec/18/technology-prevent-waste-food-developing-countries.

[11]Jonathan Rees, *Refrigeration Nation,* Baltimore: Johns Hopkins University Press, 2013, Kindle edition, loc. 1736-1767.

[12, 13]Julian Parfitt et al, "Food Waste Within Food Supply Chains: Quantification and Potential for Change to 2050," Philosophical Transactions of The Royal Society, September 27, 2010, http://rstb.royalsocietypublishing.org/content/365/1554/3065.full.

[14]*The Role of Refrigeration in Worldwide Nutrition,* "5th Informatory Note on Refrigeration and Food," Intergovernmental Organization for the Development of Refrigeration, International Institute of Refrigeration, Paris, June 2009, http://www.iifiir.org/userfiles/file/publications/notes/NoteFood_05_EN.pdf.

[22, 23]Nicola Twilley, "What Do Chinese Dumplings Have to Do with Global Warming," *The New York Times,* June 25, 2014, http://www.nytimes.com/2014/07/27/magazine/what-do-chinese-dumplings-have-to-do-with-global-warming.html?_r=0.

[25]Pravin Palande, "How Coldex Logistics is Climbing Up the Food Chain," *Forbes India,* May 17, 2015, http://forbesindia.com/article/big-bet/how-coldex-logistics-is-climbing-up-the-food-chain/39825/1.

[26]Gaurav Jain, correspondence with the authors, May 29, 2015.

[27, 28]Melanie Lee, "E-commerce Heats Up Cold-Chain Logistics in China," *Internet Retailer,* October 12, 2014, https://www.internetretailer.com/2014/10/12/e-commerce-heats-cold-chain-logistics-china.

[29]T. Reardon and J.A. Berdegue, "The Rapid Rise of Supermarkets in Latin America: Challenges and Opportunities for Development," *Development Policy Review,* 20, 371-388.

[30]"IARW Global Cold Storage Capacity Report Shows Strong Worldwide Growth," Global Cold Chain Alliance, December 23, 2014, http://www.gcca.org/press-releases/iarw-global-cold-storage-capacity-report-shows-strong-worldwide-growth/.

[31]Based on 2014 population estimates of 318.9 million residents in the U.S., 1,287.0 million in India and 1,393.8 million in China. Sources: World Bank, U.S. Census Bureau and the United Nations.

[32]Jason Turek, interview with the authors, May 11, 2015

[33]Frank Turek Jr., interview with the authors, May 11, 2015

[34]"Arab countries" were defined in the report as all members of the League of Arab States.

[35, 36, 37]"The Grain Chain: Food Security and Managing Wheat Imports in Arab Countries," The World Bank, 2012, http://www-wds.worldbank.org/external/default/WDSContentServer/WDSP/IB/2012/04/17/000333038_20120417011859/Rendered/PDF/680750WP0P11730hainpubENG4013012web.pdf.

[38]"Missing Food: The Case of Postharvest Grain Losses in Sub-Saharan Africa," The World Bank, 2011, http://siteresources.worldbank.org/INTARD/Resources/MissingFoods10_web.pdf.

[39]Hippolyte Affognon et al, "Unpacking Postharvest Losses in Sub-Saharan Africa: A Meta-Analysis," *World Development*, Vol. 66, 2015, pp. 49-68, http://www.sciencedirect.com/science/article/pii/S0305750X14002307.

[40]Refrigeration Industry Leaders Organize Global Food Cold Chain Council," September 23, 2014, http://www.prnewswire.com/news-releases/refrigeration-industry-leaders-organize-global-food-cold-chain-council-276595941.html.

[41]Kevin J. Fay, interview with the authors, May 15, 2015.

[42]CocaCola Enterprises, accessed 2014, http://www.cokecce.com/corporate-responsibility-sustainability/energy-and-climate-change.

[43]*Refrigerants, Naturally!,* http://www.refrigerantsnaturally.com/.

[44]S.J. James and C. James, "The Food Cold-Chain and Climate Change," Food Research International, http://ucanr.edu/datastoreFiles/608-150.pdf, 2010, 1950.

CHAPTER 5: A MOUNTAIN OF WASTED FOOD

A MOUNTAIN OF WASTED FOOD CAN TELL A THOUSAND TALES.

Take the story of Hong Kong, for example. Located off of China's south coast, this beautiful city is home to a deep natural harbor and one of the world's most celebrated skylines. It is also one of the most intensely urbanized communities on the planet. Some 7 million residents occupy just over 400 square miles; by comparison, the London metropolitan region has about twice the population but almost nine times the land mass.

Managing waste effectively plays an important role in maintaining Hong Kong's quality of life.

The city generates a half-kilo or roughly 1 pound per person of food waste daily. About two-thirds of this comes from households and a third from the city's schools, hotels, grocers and restaurants.[1] Some fresh and edible castoffs are used to feed the hungry through local food recycling programs. Some food waste ends up as compost in local farms.

However, far too much ends up in one of the city's three gigantic landfills. Food accounts for a third of all solid waste in Hong Kong—over 3,000 metric tons daily.[2] This has doubled in the last five years, straining landfills already bursting at the seams. A mountain of food waste threatens the very livability of this beautiful and historic city.

Or consider what happened in Florida, a very different kind of story. In 2010, a hard freeze collapsed the timetable for harvesting strawberries in southwest Florida. The market was suddenly flooded with perfect red berries. Fruit that might have fetched $1.00 per pound was selling for $.25 per pound. Most farmers stood to lose money if they continued picking and shipping their crop. Some allowed residents and food banks to gather berries for free. Others, worried about injuries and liability, left the fruit to decompose. Without programs in place to distribute the excess product, it was a no-win situation for everyone. As one unhappy farmer said, "We don't mean to hurt anybody... We're just trying to make a living like everyone else is."[3]

This is all too common a story in the U.S., where some 7 percent of planted fields go unharvested every year due to changes in demand or pricing.[4] Even with modern agriculture and a wealthy consumer, nutritious produce is still at risk of being discarded in our mountain of waste.

And then there's the tale of Delhi, India, home to one of Asia's largest produce wholesale hubs. Each day dozens of trucks roll into the busy market at Azadpur Subzi Mandi. Each is full of produce waiting to be unloaded and repacked in smaller trucks for local delivery. Much of this activity is done in high humidity under a scorching sun. Some 40 percent of all fruit and produce that passes through this important hub will rot before it reaches consumers.[5]

These are just three of the tales told by our mountain of wasted food, three ways in which good food goes uneaten. In Hong Kong, waste looks like enormous, overflowing landfills. In the United States, it appears as fields of flawless, unpicked strawberries. In India, it's a giant produce hub without enough cold storage warehousing and transportation to protect fruits and vegetables heading to market.

LOSS AND WASTE BY THE NUMBERS

Careful definitions may seem absurd when we're staring at a mountain of rotting food in a landfill. But in trying to actually solve the problem of loss and waste, good definitions can yield important clues.

Consequently, for the precise definition in this chapter we use *food loss* to describe the decrease of food mass during agricultural production and post-harvest handling and storage. Sometimes this is referred to as "upstream" in the food supply chain. *Food waste* is all the food discarded by retailers and consumers who are "downstream" in the food supply chain. Sound measurement is also important, so let's start from the very top of the food mountain. Annual global agricultural production for food and non-food uses is about 6 billion metric tons. Of that, some 2 billion are used for industrial and non-edible purposes.[6] That leaves 4 billion metric tons of edible food, fit for human consumption. From that splendid, nutritious supply—enough to feed the entire world—we

AN ELEPHANT WEIGHS ABOUT **1** **METRIC TON**

NOW IMAGINE **1.3** **BILLION HEALTHY ELEPHANTS** HAPPILY STANDING ON EACH OTHER IN ONE PILE. THAT'S WHAT WE'RE LOSING AND WASTING FROM THE FOOD SUPPLY CHAIN EACH YEAR

lose and waste about one-third, or 1.3 billion metric tons.[7]

Here's a different way to think about it. An elephant weighs about 1 metric ton. Now imagine 1.3 billion healthy elephants happily standing on each other in one pile. That's what we're losing and wasting from the food supply chain each year. In monetary terms, the most recent global food loss estimate is a staggering $1 trillion in retail value,[8] or about twice the gross domestic product of Norway.[9]

FINDING THE FOOD-WASTE TROUBLE SPOTS

Tackling the problem of food loss and waste can seem overwhelming. One good way to start is by identifying the most pressing trouble spots in the global food supply chain.

With its huge population and relatively undeveloped cold chain, it's no surprise that Asia suffers most from food loss and waste. The pain is shared by both developed and undeveloped regions on this huge continent. Industrialized Asia absorbs 28 percent and South and Southeast Asia 22 percent—50 percent of total global food loss.[10] Asia grows and consumes more than 50 percent of the world's vegetables, a category which represents some 15 percent of total global loss. Cereals add another 15 percent. Asia and Europe together also lose massive amounts of starchy

roots (like potatoes, yams, beets and carrots) amounting to 13 percent of global food loss.[11] Clearly, if we want to tackle food loss, we need to look for ways to expand and enhance Asia's embryonic cold chain.

Another trouble spot is fruit loss in South and

Southeast Asia, as well as Latin America. This totals about 7 percent of global loss. It includes our friend the banana, but also a vast variety of fruits that are dependent on the same undeveloped cold chain through which vegetables and starchy roots travel. Enhancing the cold chain for one perishable category will improve it for all.

In seeking solutions, there's another useful way to slice the numbers. It's also a way that makes clear that the issue of food loss and waste is truly global and not the responsibility of any one region. What does food loss look like per person? On a per capita basis, Europe, North America, Oceania and Industrialized Asia waste between 300 and 340 kg of food per year. South and Southeast Asia, despite high absolute waste, have among the smallest per capita at 160 kg.[12]

In addition, in medium- and high-income regions, most waste occurs at the end of the supply chain when food is discarded by consumers and retailers. This means that energy inputs such as harvesting, transportation and packaging are embodied in the food. For example, if we must waste a tomato, it's relatively better to have it decompose in the field rather than pick, clean, pack, cool, ship and display it at retail, only to have it thrown out by a consumer.

Food waste at the consumer and retail level can be as much as 39 percent of all waste in middle- and high-income countries, but as low as 4 percent in low-income countries.[13] All told, developed countries can play a substantial

FOOD WASTE:

1/3 CONSUMERS

2/3 PRODUCTION AND DISTRIBUTION

role in curbing food waste.

The United States discards 133 billion pounds of food annually, or about 31 percent of its total supply. Forty-three billion pounds are tossed at retail and 90 billion by the consumer.[14] That's $161.6 billion of losses, with meat, poultry and fish (30 percent), vegetables (19 percent) and dairy (17 percent) leading the way.[15] In a culture that some-times seems obsessed with calories, Americans discard some 141 trillion annually.[16]

ANOTHER SPIN OF THE NUMBERS

As we seek solutions, there's another significant spin of the numbers. Of all lost and wasted food, 85 percent comes from plant products and 15 percent from animals. Vegetables and fruits account for 40 percent of all waste, cereals for 25 percent, and starchy roots for 19 percent.[17]

In fact, while Asian produce is the most pressing problem globally in sheer volume, vegetables, fruits and starchy roots contribute to high levels of food loss and waste in every region of the world.[18] The products that can supply the micronutrients needed to solve the massive problem of hidden hunger are the ones most susceptible to loss.

In the U.K., 11 types of fruits and vegetables were mapped across five critical points in their supply chains, including field, grading, storage, packing and retail. Perhaps surprisingly, different products had enormously different "resource maps." For example, potatoes are a food that experienced very little loss in the field, in storage or at retail. But up to 13 percent of potatoes can be intentionally discarded in the field because they are imperfect, and another 25 percent when graded again in storage. In the case of potatoes, waste is a function of size, shape and cosmetics. Conversely, apples do well in storage, packing and at retail, but can suffer catastrophic losses in the field (up to 100 percent due to weather) and at grading, where water loss, firmness and bruises are all common.[19] This suggests that cold chains must be not only robust, but flexible enough to adapt to the idiosyncrasies of different perishable products.

A COMPLICATED PUZZLE

There are immense complexities buried beneath the mountain of global food waste. Produce, grains and starchy roots can be lost to inferior seed and fertilizer, drought or flood, poor crop rotation, harvesting practices or inferior packaging. Sometimes bad roads and the absence of electricity contribute. Weather and disease are constant threats. Supply and demand lead to price shocks, causing food to go unharvested or making it unaffordable. Perishables ranging from produce to meat and dairy require adequate cold storage facilities and transport. Supply chains need expertise and strong communications linking upstream and downstream activities. Even when the journey from "farm to fork" is flawless, restaurants must still serve wisely, grocers sell wisely and shoppers plan and consume wisely.

Our stories from the mountain also reinforce the global nature of food loss. While Asia represents 50 percent of loss, it also represents 60 percent of the global population. That means the remaining 40 percent of humankind is responsible for half of the problem—and therefore being more wasteful on a per person basis. Likewise, there are leaks along the entire food chain; upstream losses by weight total 54 percent, while downstream accounts for 46 percent.[20] In terms of climate impact, "downstream" consumers are responsible for 37 percent of all carbon emissions.

All loss hurts. And the pain of food loss and waste is spread geographically, up and down the supply chain and across a wide variety of food products. A banana can be lost to weather on a plantation in Central America, to heat on a ship in the Atlantic, or to cosmetics in a kitchen in Paris.

Tales from the mountain also suggest that there are two very different kinds of problems associated with food loss and waste. One is structural in nature: bad weather, poor roads, improper packaging and an inadequately refrigerated distribution system. Many of these issues can be addressed through careful planning, political will and sufficient investment. And then there are problems that are economic and cultural in nature, powerful forces almost built into the system. Food too expensive to be purchased will rot in the warehouse. Food too unprofitable to harvest will be lost in the field. Meal servings that are twice what a person can eat will be partially discarded. A perfectly edible apple with harmless spots or a misshapen carrot might be tossed in a landfill if there are cheap and perfect alternatives. The elements of supply and demand, pricing, tradition and culture all play an important role in food loss and waste.

Most of all, tales from our mountain of food make clear that there are challenges and opportunities enough for the entire global community. It takes a village to build a mountain—and it will take our entire global village to level it.

The stakes are exceedingly high. If we are not successful in meeting these challenges and continue to waste one-third of all food, hundreds of millions of people will remain hungry and billions malnourished.

[1]Katie Hunt, "Banquet-Loving Hong Kong Grapples With Mountains of Food Waste," BBC News, January 2, 2013, http://www.bbc.com/news/business-20807819.
[2]"Food Waste Stats," Feeding Hong Kong, accessed 2015, http://feedinghk.org/hunger-stats/.
[3]"Wasting Strawberry Fields," ABC News, March 2010, http://abcnews.go.com/WNT/video/wasting-strawberry-fields-10220551.
[4]C.G. Winkworth-Smith et al, "The Potential Value of Reducing Global Food Loss," The University of Nottingham, Division of Food Sciences, School of Biosciences, March 2015, 21.
[5]Amy Kazmin, "India Tackles Supply Chain to Cut Food Waste," April 11, 2014, http://www.ft.com/cms/s/2/c1f2856e-a518-11e3-8988-00144feab7de.html#axzz3UvtEm5TU.

[6]"Food Losses and Waste in the Context of Sustainable Food Systems," a report by the High Level Panel of Experts on Food Security and Nutrition of the Committee on World Food Security, Rome, 2014, http://www.fao.org/3/a-i3901e.pdf, 11.

[7]These numbers are from 2007. *Food Wastage Footprint: Impacts of Natural Resources, Technical Report* (Working Document), FAO, 3.

[8]*Food Wastage Footprint: Full-cost Accounting,* Final Report, FAO, 2014, http://www.fao.org/3/a-i3991e.pdf, 7-8.

[9]"List of Countries by GDP (2013)," *Wikipedia,* http://en.wikipedia.org/wikiList_of_countries_by_GDP_%28nominal%29.

[10, 11, 12, 13]*Food Wastage Footprint: Impacts of Natural Resources, Technical Report* (Working Document), FAO, 104, 112, 110, 108.

[14, 15, 16]Numbers based on 2010 estimates. Jean C. Buzby et al, "The Estimated Amount, Value, and Calories of Postharvest Food Losses at the Retail and Consumer Levels in the United States," United States Department of Agriculture report summary of the Economic Research Service, February 2014, http://www.ers.usda.gov/publications/eib-economic-information-bulletin/eib121.aspx.

[17, 18, 20]*Food Wastage Footprint: Impacts of Natural Resources, Technical Report* (Working Document), FAO, 103, 107, 104.

[19]"Fruit and Vegetable Resource Maps," WRAP, June 2011, http://www.wrap.org.uk/sites/files/wrap/Resource_Map_Fruit_and_Veg_final_6_june_2011.fc479c40.10854.pdf.

CHAPTER 6: FOOD WASTE AND CLIMATE CHANGE

WE SHOULD REALLY ALL BE FREEZING.

The basic laws of physics tell us that it should be bitterly cold on the surface of the Earth. There's just not enough warming sunlight across 93 million miles of space to melt ice. Our sun is simply too far away.

Of course, our basic senses tell us something entirely different. We can walk in the sunshine, lie on the beach, work up a sweat—and harvest food across wide regions of the planet. Our comfort and productivity are the result of "greenhouse" gas molecules in Earth's lower atmosphere. These molecules capture and hold infrared radiation from the Earth's surface. In fact, it's thanks to the two most important GHGs, water vapor and CO_2,[1] that humankind is here at all.

GREENHOUSE GASES: TOO MUCH OF A GOOD THING

Until about 1950, nobody talked much about GHGs, global warming or climate change. People generally understood that the Earth had warmed over the previous 75 years, but the study of climate was rudimentary and sometimes contradictory. It was only in the 1950s that precise measurements confirmed a buildup of CO_2 in the atmosphere. Other gases were also accumulating, some degrading the planet's ozone layer. By the 1970s a handful of scientists grew alarmed. Then the summer of 1988 brought heat-related deaths, drought, raging forest fires and livestock dying in fields. In some parts of the world it was the hottest summer on record to that time. That's when the general public first became broadly engaged in the climate change debate. By 2000 predictions of unprecedented global warming were common, generating debate over the extent and causes of climate change.

Today, global warming and its causes are not without controversy. But the United Nations' Intergovernmental Panel on Climate Change (IPCC) has come to reflect the consensus of thousands of scientists and experts. Its work is supported by the governments of more than 120 countries. When the IPCC speaks, people listen. The panel's 2014 "Summary for Policymakers,"[2] subject to line-by-line approval by delegates, concluded the following:

- Human influence on the climate system is clear, and recent anthropogenic [human-made] emissions of greenhouse gases are the highest in history. This has led to atmospheric concentrations in CO_2, methane and nitrous oxide that are unprecedented in the last 800,000 years.
- Continued emission of greenhouse gases will cause further warming and long-lasting changes in all components of the climate system.
- Risks are unevenly distributed and are generally greater for disadvantaged people and communities.

The IPCC listed among its mitigation strategies improved access to nutrition, improved infrastructure, changes in cropping and livestock practices, maintaining wetlands and watersheds, protecting genetic diversity, and establishing new food storage and preservation facilities.

That sounds remarkably like a recipe for fixing the

global food system. The IPCC's work underscores the fact that food production, distribution, consumption and waste are inextricably bound up with one of the most pressing issues of our time.

MAKING THE CONNECTION: FOOD WASTE AND CLIMATE CHANGE

Determining the carbon footprint of a food product is tricky business. It involves measuring the total amount of GHGs emitted throughout the entire life of that product, including planting, growing and breeding, transport, processing, packaging, consumption and waste. It's a long, intricate chain that in total can reveal for each product or category a single carbon footprint value measured in CO_2 equivalents.

In the case of most food products, it's methane and nitrous oxide that have the greatest impact on the environment. Methane is rated 25 times and nitrous oxide 298 times more powerful than CO_2.[3] Methane is especially prevalent in the enteric fermentation of ruminants (or, to be blunt, the belching and flatulence produced by animals like cows). Methane is also formed when food degrades in a landfill. Nitrous oxide is common in fertilizer and therefore associated with all vegetables and animal feeds. For purposes of comparison across all sources of GHG—from autos and grocery stores to dairy farms and banana plantations—the impact of methane and nitrous oxide is converted to the same standard CO_2 equivalent.

Emissions caused by changes in land use are not yet included in current studies. The "competition for land" we discussed in a prior chapter can have a material impact on the environment. For example, current estimates for greenhouse gas emissions from global agriculture range from 11 percent to 15 percent of worldwide carbon emissions. The U.S. Environmental Protection Agency estimates 14 percent,[4] which includes emissions from equipment, soil management and livestock management. However, the United Nations Conference on Trade and Development suggests that global agriculture contributes between 43 percent and 57 percent[5] of GHGs once land use, deforestation, transport, food processing, packaging, and sale of agricultural products are included. Whatever the number, reducing agriculture's global footprint from harvest to waste is a compelling way to reduce carbon emissions.

CLIMATE IMPACT BY THE NUMBERS

Let's again climb to the top of our mountain of food waste. The view we're seeking this time is of something called a "sustainable food system." This is one that not only delivers food security and nutrition today, but also leaves intact the economic, social and environmental means required to generate food security and nutrition for future generations.[6] In other words, it's not enough just to feed today's 7 billion people. We need to leave enough good land, clean air, freshwater and biodiversity to sustainably feed the more than 9

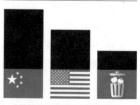

IF FOOD WASTE WERE A COUNTRY IT WOULD BE THE THIRD LARGEST EMITTER OF GREENHOUSE GASES

billion expected by 2050.

With that in mind, the current environmental numbers are challenging at best. When we convert all of the activity required to create our mountain of waste, gathered from all of the food supply chains worldwide, we find that global food loss and waste generate 3.3 billion metric tons of CO_2 equivalent annually.[7] In a single year, all of the good food we never get to consume equals 60 times the current carbon footprint of New York City.[8]

These are extraordinary numbers, and it takes a moment for the environmental dimension and impact of food waste to register. This total—3.3 billion metric tons of CO_2—is more than two times the GHG emissions from road transportation in the U.S.[9] If food loss and waste were its own country, it would rank as the third highest GHG emitter after China and the U.S.[10] Sadly, if the 805 million hungry people in the world suffering from this profligate waste were their own nation, it would also be the third largest country in the world.

Let's say that again: In both hunger and climate impact, food loss and waste is the world's third largest country.

The major contributors to the carbon footprint of food waste are cereals (34 percent), meat (21 percent) and vegetables (21 percent). Taken together, these three categories account for more than 60 percent of the carbon footprint in every region around the world.[11] Cereals are impacted by emissions from the use of fertilizer, harvest and drying equipment. Rice is a major crop of Asia, and when grown wet in paddies can produce methane. For starchy roots the equation is reversed: Their carbon footprint is less than a third of their 19 percent waste estimate. Fisheries represent a relatively small food category but do impact the environment through the use of ship fuel and refrigerants.

FINDING THE CARBON-FOOTPRINT TROUBLE SPOTS

Just as we analyzed food waste, it's worth slicing the numbers for food waste's carbon footprint by geography, food category and supply chain segment to determine where focus and investment might make the greatest improvement.

In terms of location, the major contributors to the carbon footprint are Industrialized Asia (34 percent) and South and Southeast Asia (21 percent).[12] At 55 percent for all of Asia, this is slightly more than its waste percentage. GHG emissions are also relatively high in North America and Europe because diets are so rich in meat. In environmental terms, a small amount of wasted meat can have a substantial impact compared with many types of lower-energy perishables.

Looking at the carbon footprint of waste across the supply chain highlights the fact that the more value added to a product, the greater its embodied emissions. In some middle- and high-income regions, downstream GHG emissions can be three times higher than upstream.[13] In those same nations, a combination of abundant meats and processed foods, a robust supply chain that moves product efficiently downstream, and high consumer waste are an Achilles heel. Europe, North America, Oceania and Industrialized Asia waste 700 to 900 kg CO_2 equivalent per capita. Sub-Saharan Africa has the world's smallest per capital carbon footprint at 180 kg CO_2 equivalent.[14]

The contrast between meat and vegetable diets was captured in a British study of some 65,000 U.K. residents. Average GHG emissions associated with a "high meat-eater" diet were 1.85 times greater than vegetarians and nearly 2.5 times higher than vegans.[15] And the consumption of meat continues to grow. Across the Atlantic, Americans consumed 41 percent more meat in 2000 than they had in 1950.[16]

When we combine these various findings, the global trouble spots for carbon footprint appear. Asia holds all four of the top spots, reflecting GHG emissions for cereals (1), vegetables (3) and meat (4) in Industrialized Asia, and cereals (2) in South and Southeast Asia.[17] European vegetables (5), meat (8) and cereals (9) make up three of the remaining six critical problem areas.[18] When we factor in a look across the supply chain, the trouble spots for carbon footprint are spread almost evenly between upstream and downstream, with cereals, vegetables and meat dominant.

A CALL TO ACTION: THE MOST HELPFUL VIEW

As we ponder how best to eliminate waste and feed the hungry from today's global food supply chain, we also need to create tomorrow's sustainable food system. In other words, the most effective solutions will be ones that feed the hungry and protect the planet.

Here the numbers speak volumes. FAO estimates[19] pinpoint the following five carbon-footprint trouble spots:
1. Cereals in Asia
2. Vegetables in Asia

3. Meat in Industrialized Asia

4. Vegetables, meat and cereal in Europe

5. Meat in North America and Oceania

There appears to be no single greater opportunity for both reducing hunger and climate impact than enhancing the cold chain in Industrialized Asia.

If today's food system remains unchanged and global agriculture is forced to increase production by 70 percent to satisfy world demand in 2050, GHG emissions could grow by 30 percent.[20] The laws of physics suggests that things could get awfully warm at the top of our mountain.

The good news is that reducing food loss and waste will feed the hungry and decrease global warming. The better news is, that's just the start.

[1]Spencer Weart, "Introduction: A Hyperlinked History of Climate Change Science," *The Discovery of Global Warming,* Spencer Weart and the American Institute of Physics, February 2015, http://www.aip.org/history/climate/index.htm.

[2]"Climate Change 2014 Synthesis Report Summary for Policymakers," accessed 2015, http://www.ipcc.ch/pdf/assessment-report/ar5/syr/AR5_SYR_FINAL_SPM.pdf.

[3]These are values used by the FAO. The USDA uses slightly different global warming potentials of 21 for CH4 and 310 for N2o.

[4]"Global Greenhouse Gas Emissions Data," United States Environmental Protection Agency, accessed 2015, http://www.epa.gov/climatechange/ghgemissions/global.html#two.

[5]"Wake Up Before It's Too Late," United Nations Conference on Trade and Development, Trade and Environment Review 2013, 20, http://unctad.org/en/publicationslibrary/ditcted2012d3_en.pdf,

[6]"Food Losses and Waste in the Context of Sustainable Food Systems," a report by the High Level Panel of Experts on Food Security and Nutrition of the Committee on World Food Security, Rome, 2014, http://www.fao.org/3/a-i3901e.pdf, 31.

[7]These numbers are based on 2007 estimates. *Food Wastage Footprint: Impacts of Natural Resources, Technical Report* (Working Document), FAO, 116. More recent updates show 3.5 billion metric tons of CO_2 equivalent. *See Food Wastage Footprint: Full-cost Accounting,* Final Report, FAO, 2014, http://www.fao.org/3/a-i3991e.pdf, 7.

[8]"New York City's Carbon Emissions – In Real Time," *Carbon Visuals,* 2015, http://www.carbonvisuals .com/work/new-yorks-carbon-emissions-in-real-time. The estimate for 2010 was 54 million metric tons. Divided into 3.3 billion and rounded, that's 60.

[9, 10, 11, 12, 13, 14, 17, 18, 19]*Food Wastage Footprint: Impacts of Natural Resources, Technical Report* (Working Document), FAO, http://www.fao.org/3/a-ar429e.pdf, 132, 3, 124, 117, 120, 126, 128, 129.

[15]Peter Scarborough et al, "Dietary Greenhouse Gas Emissions of Meat-eaters, Fish-eaters, Vegetarians and Vegans in the UK," British Heart Foundation Centre on Population Approaches for Non-Communicable Disease Prevention, Nuffield Department of Population Health, University of Oxford, June 11, 2014, http://download.springer.com/static/pdf/694/art%253A10.1007%25 2Fs10584-014-1169-1.pdf?auth66=1427048285_a61e173002b1b8d9aa7f7b8b900a5e13&ext=.pdf.

[16]Dean Ornish, "The Myth of High-Protein Diets," *The New York Times,* March 23, 2015, http://www.nytimes.com/2015/03/23/opinion/the-myth-of-high-protein-diets.html.

[20]*FAO Food and Nutrition in Numbers 2014,* http://www.fao.org/publications/card/en/c/9f31999d-be2d-4f20-a645-a849dd84a03e/, 50.

CHAPTER 7: WASTED FOOD, WASTED WATER

DON'T BE FOOLED BY THE FACT THAT NEARLY THREE-QUARTERS OF EARTH'S SURFACE IS COVERED BY WATER.

Just as we found with land, this "big picture" look at our planet is comforting but misleading. It turns out that only about 2.5 percent of the Earth's water is freshwater, much of it locked in glaciers, ice caps and permafrost. Just 1.3 percent of the world's total remaining freshwater is available as surface water in rivers, lakes, ice and snow.[1]

1.3% OF THE WORLD'S **FRESHWATER** IS AVAILABLE AS SURFACE WATER IN **RIVERS, LAKES, ICE AND SNOW**

That makes freshwater, seemingly so plentiful, one of humankind's most precious commodities.

A report issued by the United Nations for World Water Day in 2015 concluded that demand for water globally will increase 55 percent in the next 15 years. If nothing else changes, only 60 percent of the world's water needs will be met in 2030. Seven hundred million people could be displaced from their land.[2] The women of sub-Saharan Africa already spend 40 billion hours a year doing nothing but collecting water so that their families can survive.[3]

Agriculture, ecosystems, cities, economies and global health all depend on humankind having access to a sustainable supply of clean freshwater.

A SLOW-MOVING NATURAL DISASTER

Unfortunately, the crisis is already upon us. Today, one in six people around the world lacks safe drinking water.[4] Thirty-seven countries are currently under water stress due to inadequate internal freshwater resources or lack of regional cooperation.[5] More than 10 percent of people worldwide consume foods irrigated with wastewater that can contain chemicals or disease-causing organisms.[6] Stories of drought and hardship are abundant.

For example, the Yellow River—China's "cradle of civilization"—can today dry up from overconsumption before it reaches the sea.[7] That places additional stress on a country where 21 percent of the available surface water is already considered unfit for agriculture. Central Asia's Aral Sea was once one of the largest lakes in the world. Depleted by irrigation projects, it is now only a fraction of its former size. In the American South, the Rio Grande stretches 1,900 miles to the Gulf of Mexico. Today, in places like El Paso, Texas, it slows to a trickle. One estimate forecasts that this once mighty river could lose a third of its water by the end of this century.[8] In Chainat Province in Thailand, farmers are facing the worst drought in decades, forcing a shift from rice to less profitable arid crops that can survive dry conditions. Likewise, farmers in India's Karnataka state are switching from rice production to millets in an effort to persevere in the face of a relentless water crisis.[9] Food security in both regions is plummeting.[10]

In sub-Saharan Africa, 329 million people lack access to improved water supplies and 640 million do not have access to improved sanitation facilities. While rain-fed crops like maize, millet and tubers can be grown, more nutritious crops often require irrigation. Poverty and lack of investment mean that only 6 percent of the cultivated area in the region is irrigated. The inability to harness the water resources at hand contributes to hunger and malnutrition.[11]

Iran is an example of a country facing another kind of perfect storm. The nation has one-third the global average of annual rainfall and three times the evaporation rate, a growing population, and a 20 percent decline in rainfall in the last 20 years. The country's Lake Urmia, once designated a biosphere, has shrunk 90 percent in the last decade. Windstorms now spread exposed salt that could destroy farmlands in 10 of Iran's 31 provinces.[12]

In some regions prolonged drought creates desertification, a condition where once fertile land becomes desert. The U.N. describes desertification as a "silent, invisible crisis that is destabilizing communities on a global scale."[13] That's precisely what's happening to the land around the Aral Sea and Lake Urmia, and in vast tracts across Asia and Africa. Unlike a seasonal drought, desertification permanently ends all agricultural activity and intensifies poverty and food insecurity. More than 1.5 billion people around the world already depend on degraded land. Some 12 million formerly productive hectares become barren every year. Without change, crop yields in some African countries could fall by 50 percent due to desertification.[14]

Developed countries are not immune to drought. The Colorado River Basin, upon which 40 million Americans depend, lost almost 16 cubic miles of water between 2004 and 2013. This is twice the amount of the country's largest reservoir at Lake Mead.[15] In California, what's been called an "epic drought" resulted in increased wildfires and agricultural losses of more than $2.2 billion in 2014 alone.[16]

Water can threaten us in two ways. A flood poses an immediate danger and people know to react decisively. However, long-term drought is what one expert has called a "slow-moving natural disaster,"[17] gradual but every bit as catastrophic. In fact, our life-giving water also turns out to be one of the deadliest of our necessities: Between 1991 and 2000, more than 665,000 people died in natural disasters, 90 percent of which were water-related.[18]

Water is also becoming one of our most expensive commodities. Historically, governments around the globe have spent about $45 billion annually to ensure adequate water supplies for their citizens. Forecasts suggest that number could increase to $200 billion by 2030.[19]

WHEN WE WASTE FOOD, WE WASTE WATER

The competition for water is not unlike the struggle for land. Farms, cities, industry and conservation areas all require sustainable supplies of freshwater. But unlike the battle for land, there is one decisive winner: Agriculture consumes 70 percent of our global water withdrawals, with just 20 percent used by industry and 10 percent for domestic use.[20] As the global population grows—and without some fundamental change in our actions—agriculture is forecast to require about two-thirds of the additional demand for freshwater by 2030.[21] This is simply unsustainable.

When we consider ways to protect our fragile water resources, we need to look first and foremost at the global food supply chain. California provides one good example. The state produces nearly half of all U.S. fruits, vegetables and nuts from the very areas hardest hit by drought. Monterey County alone produces about half of the country's lettuce and broccoli.[22]

Now imagine a consumer rummaging around in the back of his refrigerator's vegetable drawer only to find a forgotten head of broccoli, now yellow and unappetizing. He drops it in the trash. No big deal, right?

But wait: Fresh broccoli is about 91 percent water, and that's just the start. It actually takes a farmer about 5.4 gallons of water[23] to grow that single head of broccoli. Just as each food product has an embedded carbon footprint, it also has a quantity of embedded freshwater from its journey along the food supply chain. In fact, a single person blessed with a healthy, nutritious diet will drink up to a gallon of water per day but "eat"

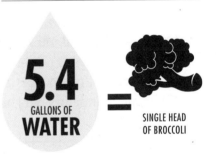

up to 1,300 gallons of embedded freshwater in his food.[24]

The farmers in California consume 80 percent of the state's developed water supply.[25] Facing a drought that began in 2011, some have seen their traditional surface allocations radically reduced or disappear entirely. Aggressive pumping of groundwater has contributed to a statewide drop of more than 12 million acre-feet of water annually since the start of the drought. In some parts of the state's Central Valley, land is sinking by more than a foot a year.[26] A mandatory 25 percent reduction in water use announced by the governor in April 2015 targeted lawns, golf courses and commercial entities. Four hundred local water supply agencies were asked to determine how best to achieve that goal.

California's farmers were initially exempted. However, they have already felt the impact: the state's $40 billion agriculture industry lost 400,000 farm acres and nearly 20,000 jobs to reduced water resources in 2014.[27] Prolonged drought threatens more acreage, more jobs and every resident's standard of living. In June 2015, California state water officials accepted a proposal by farmers in the Sacramento-San Joaquin River Delta to voluntarily reduce their water usage by 25 percent, leaving a quarter of their land unplanted.[28] While farmers in this region own less than 10 percent of California's agricultural land, this proposal could be more broadly adopted and help stave off mandatory cuts in agricultural water usage.

Charlie Sweat is the former CEO of California's Earthbound Farm, the largest producer of organic produce and specialty salads in the U.S. "When people talk in macro terms they oversimplify the water problem in California," Sweat says. "On a macro level the state is in dire straits after four years of drought—no doubt about it. But the real issues are determined by microclimates and not evenly distributed."

The farmers in California's Central Valley, where

temperatures soar in the summer and evaporation rates are high, have been hardest hit. The types of crops grown, including tree nuts, carrots, potatoes, and tree fruits, are items that require lots of water. "Farmers will take their water allotment and they'll determine how many acres they can productively farm," Sweat says. "They will lose acreage and trees. Some farmers think they might rip out the trees and use that land for development: The problem is that if you lose the agricultural production, that acreage is not valuable for shopping malls or new homes because there are no jobs."

Farmers may also seek options that maintain productive use of their land despite reduced supplies of water. "Some farmers will explore producing other crops that require less water. So some crops may increase in volume and decrease in price while others are reduced—but with price increases," Sweat adds. "Technology and data analytics will also play a pivotal role to help conserve water and reduce waste, and to find ways to create more water reserves."

"But," Sweat concludes, "many farmers will just have to limit their acres to their water supply. Food prices are going up for sure."[29]

A SMALL TRAGEDY

As innocent a gesture as it seems, discarding that forgotten head of broccoli turns out to be a small tragedy. Imagine this happening in refrigerators all over the world. Or losing a truckload of produce exposed to the sun in a market in India. Or think about an unpicked strawberry field in Florida. In fact, we can make our way down a list of food products and begin to total up the impact of food loss and waste on humankind's freshwater resources. It takes 3.3 gallons to grow a single tomato and nearly a half-gallon to grow one strawberry.[30] More water is used by California's almond farmers annually than by all the homes and businesses in San Francisco and Los Angeles combined.[31]

Beef production is also water-intensive, significant in a world experiencing a livestock revolution. A food safety recall in 2008 of 143 million pounds of raw and frozen beef meant the loss of some 170 billion gallons of embedded freshwater—roughly equal to 20 days supply for all of New York City.[32]

Research conducted in the United Kingdom shows that more than 5 percent of the water used by households is discarded in the form of uneaten food.[33] And because much of that food is imported, more than three-quarters of the lost water has been embedded in products being shipped from countries experiencing water stress.[34] It is a hidden trade imbalance that never makes the front page news.

WATER IN WASTED FOOD > WATER USED BY ANY NATION

In a world where people are dying from lack of freshwater, these losses are tragic. Globally, lost and wasted food accounts for 250 cubic kilometers of surface and groundwater resources. This is equivalent to 38 times the annual U.S. domestic water consumption or three times the volume of Lake Geneva.[35]

In fact, the global freshwater loss related to food waste is higher than the national water usage of any country.[36] Said differently, when it comes to squandered freshwater, food waste and loss is the largest country in the world.

Nobody doubts that using freshwater to grow nutritious food is one of the very best uses of this fragile resource. Farmers in places like California argue convincingly that the 80 percent of water they consume is really water enjoyed by everyone from farm to suburb to city in the form of fresh fruits, vegetables and nuts. So the message is clear: When we lose or waste a food product, we are simply pouring freshwater down the drain.

WHERE DO WE WASTE THE MOST WATER?

As you might expect, the more populated the country, the greater its water usage. Consequently, China, India and the United States rank one through three and together account for about 38 percent of global water usage. Along with Pakistan, Australia, Uzbekistan and Turkey, they also represent the largest embedded (or "virtual") water exporters. In other words, these seven countries are shipping freshwater to other nations in the form of embedded water contained in exports. Ironically, all of these top "water exporting" countries are under some form of water stress.[37]

Some 76 percent of the virtual water that flows between countries is related to the trade of crops.[38] In this regard it's helpful perhaps to think of the global food supply chain as a massive web of hoses transporting freshwater in the form of meat, fruits, vegetables and oil crops (such as cotton and soy) between countries. When food is lost or wasted along the way, the supply chain is just one giant, leaky hose. Where a modern cold chain exists, the leaks are often patched.

Some countries depend heavily upon freshwater resources from other regions. Italy, Germany and the U.K. "import" over 60 percent of their freshwater.[39] Kuwait, Jordan, Israel, United Arab Emirates and Yemen rely on imports for over 75 percent of their water needs.[40]

A per capita look at water usage reveals that both developed and developing countries must play a part in reducing food and freshwater waste. For example, the average U.S. resident uses about 2,842 cubic meters of water annually. This is relatively high, in part because consumption of beef is about 4.5 times the global average. But Bolivia at 3,468 cubic meters annually, Niger at 3,519 and Mongolia at 3,775 all have higher per capita usage. This is because crop yields are very low. In Niger, cereal consumption is 1.4 times the global average, but the water usage per ton to grow cereals is six times the world average.[41]

Here we see a pattern we have witnessed before: Developing nations can have the greatest impact on food loss, hunger, land use, climate change, and now freshwater by focusing on upstream improvements—harvest and distribution—in the food supply chain. Developed countries need to emphasize reductions in downstream food waste.

By commodity, cereal waste contributes 52 percent of freshwater waste, followed by fruits at 18 percent. Cereals are above 22 percent in all regions and as high as 66 percent in Asia—a function of large populations and large harvest losses in wheat and rice. At the low end, wasted starchy roots contribute about 19 percent of food waste but just 2 percent of wasted freshwater. Geographically, South and Southeast Asia at 38 percent and Industrialized Asia at 20 percent contribute significantly to the loss of freshwater through food waste. Low-income regions in total are responsible for about two-thirds of wasted freshwater compared with about half the food wastage.[42]

A combination of these various measures suggests our most critical wasted-water trouble spots. First, we need to reduce the loss of cereals in Asia and North Africa, which make up 45 percent of global loss.[43] Fruits in Asia and Latin America are also heavy contributors. The impact from cereal is based on the intensity of water required to grow the commodity. From fruit, the impact on water wastage is loss after harvest.

SAVED FOOD, SAVED WATER

Let's connect the dots. When we ship food throughout the global supply chain, we ship water. In fact, nearly one-fifth of the world's freshwater is in motion between countries each year, the vast majority related to agriculture. The globe is essentially crisscrossed with large, invisible, virtual water flows related to the food supply chain. Look closely and there are any number of hidden water imbalances. This makes it evident that water is not a local or national resource, but a global and mutual dependency.

Agriculture represents the largest use of freshwater,

and we are already in trouble. The Near East and North Africa use 52 percent of their renewable water resources in irrigation. Libya, Saudi Arabia, Yemen and Egypt all use water for irrigation in greater quantities than their annual renewable water resources. In the base year of 2007, 13 countries used 40 percent of their water resources for irrigation, a level suggesting substantial water stress.[44]

Drought is making things worse and climate change is making drought worse.[45] As climate science improves, this connection becomes clearer. In simplest terms, we know that drought conditions are much more likely in low-precipitation years if high temperatures exist, all of which has a destructive impact upon the land. In 1980, about 14 percent of the Earth's land mass was classified as dry. By 2010 that number was 30 percent.[46] One estimate suggests that by 2050 about two-thirds of the global population will live in countries with water constraints, and some will be severe. These include China, India, Ethiopia, Egypt, Iran, Jordan and Pakistan.[47] Many of the countries most affected already face high levels of poverty and food insecurity.

Higher-value crops essential to healthy diets, such as vegetables and fruits, require more water per calorie than cereal crops. Meat and dairy can be five times more water-intensive than cereals.[48] The very foods we need to address global nutrition and meet consumer demand are the most water-intensive and require the greatest protection along the supply chain. Their loss and waste not only intensifies hunger, but destroys our freshwater resources. Said differently, when we stand on our 1.3 billion metric ton mountain of wasted food, we're really swimming in an ocean of squandered freshwater.

The 2015 World Economic Report on Global Risks concluded that water crises are the most serious global threat based on impact, ranking ahead of infectious diseases and weapons of mass destruction.[49] One expert commented, "In the end, we have no choice but to bring supply and demand back into balance, and the options for new supply are very limited."[50]

One obvious way to fix demand is to recognize the ocean of precious freshwater embedded in the foods we already produce, and make sure that those foods actually reach and feed humankind.

[1]FAO Food and Nutrition in Numbers 2014, http://www.fao.org/publications/card/en/c/9f31999d-be2d-4f20-a645-a849dd84a03e/, 48.

[2, 3, 5, 13, 14, 18, 46]"Desertification: The Invisible Frontline," Secretariat of the United Nations Convention to Combat Desertification, 2014, http://www.unccd.int/Lists/SiteDocumentLibrary/Publications/Desertification_The%20invisible_frontline.pdf.

[4]Amy Nordrum, "California Water Shortage: $1 Billion Plant Will Make Seawater Drinkable By End of 2015," *International Business Times*, January 27, 2015, http://www.ibtimes.com/california-water-shortage-1-billion-plant-will-make-seawater-drinkable-end-2015-1795834.

[6]*Trends and Implications of Climate Change for National and International Security*, Defense Science Board, Office of the Under Secretary of Defense for Acquisition, Technology, and Logistics, Washington, D.C., October 2011, http://www.acq.osd.mil/dsb/reports/ADA552760.pdf, 52.

[7]"8 Mighty Rivers Run Dry From Overuse," *National Geographic*, accessed 2015, http://environment.nationalgeographic.com/environment/photos/rivers-run-dry/#/freshwater-rivers-yellow-2_45150_600x450.jpg.

[8]"Once-Mighty Rio Grande is Now a Trickle," *The Orange County Register*, April 13, 2015, http://www.ocregister.com/articles/water-657768-grande-drought.html.

[9]M.T. Shiva Kumar, "Water Crisis Pushes Farmers to Switch Over to Millet Cultivation," *The Hindu,* March 20, 2015, http://www.thehindu.com/news/national/karnataka/water-crisis-pushes-farmers-to-switch-over-to-millet-cultivation/article7013630.ece.

[10]"Farmers Turn to Arid Crops to Survive," *Bangkok Post,* March 18, 2015, http://www.bangkokpost.com/news/general/499547/farmers-turn-to-arid-crops-to-survive.

[11]Laia Monènech, "How Can Reliable Water Access Contribute to Nutrition Security in Africa South of the Sahara," International Food Policy Research Institute," March 20, 2015, http://www.ifpri.org/blog/how-can-reliable-water-access-contribute-nutrition-security-africa-south-sahara.

[12]Scott Peterson, "In Thirsty Iran, a Hunt for Solutions to a Shrinking Salt Lake," *The Christian Science Monitor,* March 22, 2015, http://www.csmonitor.com/World/Middle-East/2015/0322/In-thirsty-Iran-a-hunt-for-solutions-to-a-shrinking-salt-lake-video.

[15]Dennis Dimick, "If You Think the Water Crisis Can't Get Worse, Wait Until the Aquifers Are Drained," *National Geographic,* August 21, 2014, http://news.nationalgeographic.com/news/2014/08/140819-groundwater-california-drought-aquifers-hidden-crisis/.

[16]Doyle Rice, "Intensifying California Drought Sets Off Alarms," *USA Today,* March 18, 2015, http://www.usatoday.com/story/weather/2015/03/17/california-western-drought/23953599/.

[17]Jonathan O'Callaghan, "US Faces Mega-Drought Future," *Daily Mail,* February 13, 2015, http://www.dailymail.co.uk/sciencetech/article-2952173/US-faces-mega-drought-future-Global-warming-cause-worst-dry-spells-1-000-years-claims-study.html.

[19, 21]*Resource Revolution: Meeting the World's Energy, Materials, Food, and Water Needs,"* McKinsey Global Institute, McKinsey & Company, November 2011, https://www.google.com/search?q=Resource+Revolution%3A+Meeting+the+World%E2%80%99s+Energy%2C+Materials%2C+Food%2C+and+Water+Needs%2C%E2%80%9D+McKinsey+Global+Institute%2C+McKinsey+%26+Company%2C+November+2011%2C+41.&oq=Resource+Revolution%3A+Meeting+the+World%E2%80%99s+Energy%2C+Materials%2C+Food%2C+and+Water+Needs%2C%E2%80%9D+McKinsey+Global+Institute%2C+McKinsey+%26+Company%2C+November+2011%2C+41.&aqs=chrome..69i57.404j0j4&sourceid=chrome&es_sm=93&ie=UTF-8, 47, 41.

[20]International Decade for Action "Water for Life" 2005-2015, United Nations Department of Economic and Social Affairs, accessed 2015, http://www.un.org/waterforlifedecade/index.shtml.

[22, 23]Alex Park, "It Takes How Much Water to Grow an Almond?!", *Mother Jones,* February 24, 2014, http://www.motherjones.com/environment/2014/02/wheres-californias-water-going.

[24]International Decade for Action "Water for Life" 2005-2015, United Nations Department of Economic and Social Affairs, accessed 2015, http://www.unwater.org/downloads/water_for_food.pdf

[25]Jerry Gulke, "California's Mandatory Water Restrictions: The Impact on Agriculture," *Forbes.com,* April 2, 2015, http://www.forbes.com/sites/jerrygulke/2015/04/02/californias-mandatory-water-restrictions-impact-on-agriculture/.

[26]"R.I.P. California (1850-2016): What We'll Lose and Learn From the World's First Major Water Collapse," Feelguide, March 22, 2015, http://www.feelguide.com/2015/03/22/r-i-p-california-1850-2016-what-well-lose-and-learn-from-the-worlds-first-major-water-collapse/.

[27] Darryl Fears, "As Water Runs Dry, Californians Brace For a New Way of Life," *The Washington Post,* April 4, 2015, http://www.washingtonpost.com/national/health-science/as-water-runs-dry-californians-brace-for-a-new-way-of-life/2015/04/04/f1ebb4ba-daba-11e4-b3f2-607bd612aeac_story.html.

[28]Natasha Geiling, "Farmers Agree to Water Cuts to Help California Deal With Drought," *ClimateProgress,* May 26, 2015 , http://thinkprogress.org/climate/2015/05/26/3662629/california-farmers-voluntary-cuts/.

[29]Charlie Sweat, telephone interview with the authors, April 7, 2015.

[30]Alex Park, "It Takes How Much Water to Grow an Almond?!", *Mother Jones,* February 24, 2014, http://www.motherjones.com/environment/2014/02/wheres-californias-water-going.

[31]Alex Tabarrok, "The Misallocation of Water," Marginal Revolution, March 26, 2015, http://marginalrevolution.com/marginalrevolution/2015/03/the-misallocation-of-water.html.

[32]Lundqvist, J., C. de Fraiture and D. Molden, *Saving Water: From Field to Fork—Curbing Losses and Wastage in the Food Chain,* SIWI Policy Brief, SIWI, 2008, 25.

[33, 34]Fiona Harvey, "Throwing Food Away Sends World's Scarce Water Gushing Down the Plughole," *The Guardian,* March 22, 2011, http://www.theguardian.com/environment/2011/mar/22/binning-food-wastes-lots-of-water.

[35, 36, 42, 43]*Food Wastage Footprint: Impacts of Natural Resources, Technical Report* (Working Document), FAO, http://www.fao.org/3/a-ar429e.pdf, 3, 134, 137.

[37, 38, 39, 40, 41]Arjen Y. Hoekstra and Mesfin M. Mekonnen, "The Water Footprint of Humanity," PNAS, Vol. 109, No. 9, February 28, 2012, http://www.pnas.org/content/109/9/3232.full.

[44]Nikos Alexandratos and Jelle Bruinsma, "World Agriculture Towards 2030/2050 (The 2012 Revision)," ESA Working Paper No. 12-03, Agricultural Development Economics Division, Food and Agriculture Organization of the United Nations, June 2012, http://www.fao.org/docrep/016/ap106e/ap106e.pdf, 10-12, 105.

[45]Alice Park, "The World's Water Supply Could Dip Sharply in 15 Years," *Time,* March 21, 2015, http://time.com/3753332/world-water-day-un-warning/.

[47, 48]Lundqvist, J., C. de Fraiture and D. Molden, *Saving Water: From Field to Fork—Curbing Losses and Wastage in the Food Chain,* SIWI Policy Brief, SIWI, 2008, 11.

[49]"Part 1 – Global Risks 2015: Introduction," World Economic Forum, accessed, 2015, http://reports.weforum.org/global-risks-2015/part-1-global-risks-2015/introduction/.

[50]Doyle Rice, "Intensifying California Drought Sets Off Alarms," *USA Today,* March 18, 2015, http://www.usatoday.com/story/weather/2015/03/17/california-western-drought/23953599/.

CHAPTER 8: PRETTY SOON IT'S REAL MONEY

A BILLION HERE, A BILLION THERE, AND PRETTY SOON YOU'RE TALKING ABOUT REAL MONEY.

This droll observation, attributed to American politician Everett Dirksen, describes perfectly the topic of food loss and waste. To begin, we've calculated already that there's about $1 trillion of spoiled food in our global mountain of waste. But what about the financial impact of water scarcity? Soil erosion? Greenhouse gases? What happens when the battle for land becomes real conflict? What is the cost to families fleeing farms suffering desertification? How do we measure in financial terms the profound human costs of hunger and malnutrition associated with a food system that loses one-third of everything it produces?

In 2014 the Food and Agricultural Organization of the United Nations (FAO) prepared the first full-cost accounting of food loss and waste. Their starting point was that gigantic, trillion-dollar mountain. But then they looked beyond, taking a conservative approach to assigning financial values to a range of social and environmental issues. What they concluded is that the cost of food waste is some 2.5 times greater than lost food alone—and perhaps much more than even that.

A WORD ABOUT REAL HUNGER

Measuring the financial impact of hunger and malnutrition means having an informed definition. People who have not experienced hunger firsthand tend to think in extremes. At one end, "being hungry" is that momentary sensation that comes midmorning between breakfast and a hearty lunch. At the other extreme are the devastating images of starving children, like the victims of the Ethiopian famine in 1984 or the Horn of Africa drought and famine in 2011. This is true hunger, but the kind experienced episodically and by a comparatively small percentage of the world's hungry.

The kind of hunger with which hundreds of millions suffer daily comes about through persistent undernourishment. There are enough calories to live but not enough to thrive. Hunger saps people of their energy. It leaves victims open to sickness and disease. One Kenyan farmer said, "If you're hungry, you can't sleep at night. If you have hunger, you can't plan ahead for your life. You can't think about anything except the hunger that you have."[1] Doctors who examine children experiencing hunger see fatigue, irritability, dizziness, frequent headaches, frequent colds and infections, and difficulty concentrating. Persistent hunger leads to malnutrition and stunting. Persistent hunger kills.

Even when they don't cause death, hunger and malnutrition can delay cognitive, social and emotional development. Children perform poorly in school. When nutritious foods like fruits and vegetables are missing from diets, hidden hunger can lead to severe eyesight issues and blindness (from vitamin A deficiency), and lower IQs and irreversible brain damage (from iodine deficiency in the mother). In the first two years of life, 70 percent of the brain develops; malnourishment in this period can lead to permanent damage.

Hunger casts its shadow across multiple generations. It can make impossible any escape from the very poverty that provides its foundation.

How do we measure such devastation in financial terms? One study concluded that the long-term gross domestic product (GDP) per capita growth rate in some poor countries could be increased by half a percent if dietary calories were increased by 500 kcal/day—roughly a half-cup of mixed nuts and one banana, or two cups of whole milk and a cup of chicken meat. For a subgroup of the poorest countries, GDP could grow as much as four times faster.[2] A second study that focused on several African countries concluded that the annual costs associated with child undernutrition can be as much as 16.5 percent of national GDP.[3] Other studies suggest that micronutrient deficiencies alone cost India 3 percent of its GDP while diseases related to malnourished but overweight residents comprised 13 percent of total health care expenditures in Mexico.[4]

A recent study by Bank of America Merrill Lynch concluded that hunger and undernutrition reduce the global GDP by up to 3 percent, or some $2 trillion annually.[5] A report prepared by the International Food Policy Research Institute shows that for every dollar that a government around the globe invests in nutrition to reduce stunting, it sees an average return of *16 times*—and in some countries much higher.[6]

These numbers are encouraging for the future. If we can reduce food loss and waste and keep nutritious calories from leaking out of the food supply chain, *we can raise the entire economies of countries around the world.*

That is a connection between hunger and money worth remembering.

Developed countries bear an enormous financial burden as well when their citizens suffer from food insecurity. A study published in 2011 concluded that hunger costs America $167.5 billion annually due to lost productivity, more expensive public education, avoidable health care costs, and the price of charity to keep families fed. This did not include the cost of $94 billion for the Supplemental Nutrition Assistance Program run by the federal government. The net cost of hunger in 2010 to every American family was $1,410.[7]

A "FULL COST" LOOK AT FOOD LOSS AND WASTE

The FAO's 2014 "full cost" look at food loss and waste is a smart, ambitious study that brings together the direct, environmental and social costs of our distressed food supply chain. Nobody believes the numbers are perfect. If anything, they are likely to be understated. The study's authors caution that a number of strong assumptions are required to monetize social and environmental effects. Sometimes information exists for a single country and needs to be extrapolated for others.

We also recognize that zero food loss and waste is impossible and probably not optimal. The food supply chain is hugely complex with hundreds of trade-offs. For example, it's not hard to envision a situation where building new food storage facilities to withstand typical environmental conditions is more practical than building them to protect from the very occasional extremes. These are trade-offs that may slightly degrade an individual link in the chain but still improve the entire system.

None of this really matters. Even if not exact, the valuable work done by the FAO describes the kind of enormous opportunities available to fix our broken food supply chain. In other words, we can act from these numbers to improve human welfare.

The headline is this: A current best guess for a full accounting of global food loss and waste, considered an "informed underestimate," is $2.6 trillion annually. This is roughly the GDP of France, or *twice* the annual food expenditure in the U.S.[8]

The figure to the right describes some of the broad categories measured by the study. The new trillion-dollar direct loss from food waste is updated but consistent with the $750 billion estimate used in the FAO's earlier work.[9] Social costs are calculated through measuring risk of conflict, health damages,

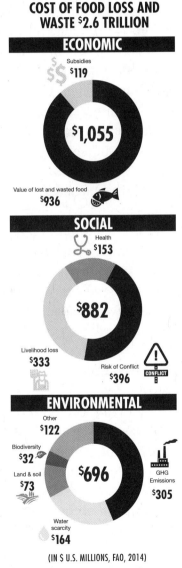

COST OF FOOD LOSS AND WASTE $2.6 TRILLION

ECONOMIC

Subsidies $119

$1,055

Value of lost and wasted food $936

SOCIAL

Health $153

$882

Livelihood loss $333

Risk of Conflict $396

ENVIRONMENTAL

Other $122

Biodiversity $32

Land & soil $73

$696

GHG Emissions $305

Water scarcity $164

(IN $ U.S. MILLIONS, FAO, 2014)

loss of livelihood, and such impacts as the acute effect of pesticides. There is no value assigned for malnutrition or hunger.

Environmental costs focus on the agricultural phase of the food supply chain, although greenhouse gas emissions are measured along the entire chain. Atmosphere, water, land and biodiversity are all measured.

At the moment, the direct cost of food loss and waste is the largest single component of its total cost. However, a full accounting suggests that every dollar of food preserved today has the multiplying financial impact of about *2.5 times*. In other words, for every dollar of wasted food saved, we receive $2.50 in health, agricultural, social and environmental benefits. Compare this with 1995 when the New York Stock Exchange[10] returned 31.3 percent, its best year ever. It would take more than three straight years of this exceptional stock performance to equal just one year's return from a dollar of food saved. As we continue to understand and measure the full costs associated with lost and wasted food, it seems possible that this multiplying effect will only grow.

WHERE IS THE FASTEST FINANCIAL PAYBACK?

We've examined global trouble spots related to food loss and waste in terms of geographic regions, commodities, carbon footprint and water scarcity. Now we're able to measure these, including land erosion, in strictly financial terms. The chart on the opposite page shows a ranking by region, with an indication of which food products might be targeted to achieve the greatest dollar improvement in each category. As with our other analyses, this reflects the impact of total population and the kinds and amounts of crops being cultivated. Whether losses occur upstream or downstream along the supply chain also has a huge influence on financial value, as does the mix of a region's diet.

THE FASTEST FINANCIAL PAYBACK: TOP OPPORTUNITIES BY CATEGORY

	$ COST		$ COST IN GHG EMISSIONS		$ COST IN WATER SCARCITY		$ COST IN LAND EROSION	
INDUSTRIALIZED ASIA	GRAINS	VEG.	GRAINS	VEG.	GRAINS	MEAT		
	MEAT	FRUITS			FRUITS			
EUROPE	GRAINS	MEAT						
	VEG.	MILK						
SOUTH & SOUTHEASTERN ASIA	MILK	GRAINS	MILK	GRAINS	GRAINS	FRUITS	MEAT	GRAIN
	FRUITS	OIL CROPS					OIL CROPS	
NORTH AMERICA AND OCEANIA							MILK	MEAT
LATIN AMERICA			MILK	VEG.			MILK	MEAT
							GRAINS	
SUB-SAHARAN AFRICA								
NORTH AFRICA & WEST/CENTRAL ASIA					PERISHABLES			

The chart above suggests a number of high-impact opportunities. For example, the "$ Cost" column shows that the greatest absolute dollar savings from reducing food waste will come from improving the food supply chains in Industrialized Asia, Europe, and South and Southeastern Asia. In particular, grains show up as costly losses in all three regions, while fruit loss is especially costly in the

Asian regions. For savings related to carbon emissions, the greatest financial impact is in Industrialized Asia, South and Southeastern Asia, and Latin America. Here, enhancing the food supply chain for grains, milk and vegetables will yield the fastest financial return. The greatest dollar impact related to water scarcity will be affected by improvements in the harvest, processing and distribution of grains in Asia and North Africa. And in terms of land erosion, the most attractive payback comes with a focus on the food supply chains for meat and milk in South and Southeastern Asia, North America and Oceania, and Latin America.

These opportunities reinforce the fact that a dollar of reduced food waste multiplies into $2.50 of savings that enhance agricultural, environmental and social resources. In short, we can save more money and feed more people while enhancing the well-being of all humankind.

While valuable, these rankings can obscure the relative magnitude of the financial impacts. In dollar terms, poor and developing nations (measured by non-Organisation for Economic Co-operation and Development (OECD) vs. OECD[11] countries) shoulder the vast majority of the global environmental costs associated with food loss and waste. This includes over 90 percent of the expense of scarce water and deforestation, over 80 percent of soil erosion, and nearly 80 percent for GHG emissions.[12] In other words, the financial burden falls on those regions least able to cope with it.

A MODEL: THE GREEN BUILDING MOVEMENT

Fixing the global food supply chain requires investment.

A study like that done by the FAO helps translate seemingly amorphous issues into financial terms. From these, attractive returns can be developed. Sometimes the humanitarian return of "doing good" is enough; certainly governments spend simply for the good of their citizens. Other times a true financial return is required to persuade people to act, especially in the private sector. The moment those two returns intersect is a moment of critical mass, when doing good and doing well align, rapidly accelerating innovation and new investment.

There is precedent for this kind of global alignment. In 1993 the U.S. Green Building Council® was formed to promote sustainability in building design, construction and operation. At the time, green investment seemed expensive and was misunderstood. "Prior to the U.S. Green Building Council," remembers Rick Fedrizzi, CEO and founding chairman, "environmental organizations and business lined up against one another. What we did at USGBC was to create a place where business could actually engage one-on-one with environmental and government organizations. By having a voice and a place at the table, some of the best ideas imaginable have come forward."[13]

Perseverance paid off. "Businesses create jobs, value and stability," Fedrizzi says. "In encouraging businesses by great examples of how they can succeed wildly with the right incentives and the right ability to create products and services that are good for the environment—and still have a rock-solid bottom line financial result—we would find our way out of the darkness that really existed for 20 or 30 years."

The tipping point for the green building movement

came when the Leadership in Energy and Environmental Design, or "LEED®" green building certification program was launched in the year 2000. By recognizing best-in-class building strategies and practices, Fedrizzi adds, "it helped people to understand how to build one of these buildings—and where the opportunity was to use their brains and innovation to create a whole better structure."

In 2002 the World Green Building Council was incorporated, and today there are more than 100 Green Councils around the world. This has created a broad-based, international movement able to measure the true cost of "business as usual" against the potential benefits of investment in sustainable buildings. "It's very difficult for anyone not to build to a LEED rating in a new construction scenario, at least in the United States today," Fedrizzi says. "LEED is used in 153 countries around the world thanks to the work of the more than 30 countries that have helped LEED develop as a global standard. It's become a game-changer. And it's starting to advance discussions relative to toxicity, water, food sourcing and social equity."

The global green building movement began as a way to protect the planet and "do the right thing." Today it has become a business imperative that drives real financial return, including significant improvements in tenant occupancy and retention with higher rents and overall building value. The vast majority of global industry professionals report specifying or using green building product. Three-quarters report that green buildings lower operating costs.[14]

"The idea that we could create an organization that actually had the ability to drive these great ideas forward in a much more meaningful way, bringing all stakeholders to the table, changed everything," Fedrizzi adds. He also sees a connection between buildings and food, especially in the context of neighborhood development that provides access to healthy foods from local markets or gardens. "Water and food may be more important than shelter," he states, noting that proper community development can encompass all three.

"I don't think a lot of people know that the problem of food waste exists," Fedrizzi says. "That of the 30 to 40 percent of all food grown and harvested with massive amounts of water and a lot of hard work—close to 40 percent of it gets thrown away. This is a crisis that is largely hidden from the public," he concludes. "If we threw away 30 to 40 percent of our oxygen I'll bet people would notice that!"

Imagine one day being able to tell a similar story about a collaborative, global effort to reduce food waste, not unlike the success of the global green building movement. We have already made a start. Thanks to the FAO and others, today we are beginning to measure and understand the full financial and environmental impact of food waste. Just like the early days of the green building movement, some players in the global food chain are already trying to "do the right thing." Increasingly we are seeing positive financial returns from such activities. Earlier we highlighted the use of silos in developing countries which reduced the loss of grain by as much as 20 percent. We also discussed

> **"IF WE THREW AWAY 30 TO 40 PERCENT OF OUR OXYGEN I'LL BET PEOPLE WOULD NOTICE THAT"**

India's Targeted Public Distribution System (TPDS), an advanced, door-to-door delivery system for food projected to save 8 to 10 percent of the country's current food subsidy.

Other examples of the emerging payback on reducing food loss include the following:

- In East Africa, milk farmers were losing about 17 percent of their product at the production and post-harvest stages, usually due to lack of adequate cooling. In response, 1,000 liter milk coolers were introduced, bringing savings of about 150 liters of milk per day per cooler. While the initial investment burden was high and electrical service could be unreliable, payback for farmers occurred in about two years.[15]

- In the Philippines, rice farmers were losing about 10 percent of their harvest due to inadequate post-harvest storage. A "Super Bag" was introduced, providing resistance to moisture, pests and fungus. Not only was rice preserved, but some quantities could be maintained into high-demand seasons and command a better price. Break-even was sometime in the second year.[16]

- In Vietnam, lack of adequate drying meant losses of quantity and quality in rice harvests. A two-stage grain-drying technology using a flash dryer, while more costly than traditional methods, improved both quality of the harvest and profits.[17]

- In Ghana, traders employ bar codes and geographic-information systems to track pine-apples from farm to port, speeding delivery and reducing spoilage.[18]

- In Switzerland, about 30 percent of carrots were lost during the processing phase, generally the result of damage at harvest, those discarded for shape and size, and overplanting to ensure quantity guarantees to processors and retailers. In a study of new carrot-sorting machines, 375 metric tons of carrots were saved from loss annually. This provided a payback sometime in the second year.[19]

One of the evolving lessons from these examples is that investing in prevention appears to have a sizable and positive financial impact, just as it does on factors like GHG emissions. And the higher the value of the product—especially fruits, vegetables, meats and foods that depend on the cold chain—the greater an impact prevention can have. A global dialogue designed to drive these important ideas, much like the collaborative work done in green building development a generation ago, could result in substantial and tangible benefits to the world's food system.

ONE MORE WAY TO LOOK AT MONEY

In many industrial settings, engineers seek a quality level that is referred to as "six sigma." This translates to a production process that has no more than 3.4 defects per million opportunities. In 2013 there were 36.4 million airline flights and 81 accidents, far better than six-sigma performance. But if airlines operated at a five-sigma quality level, there would have been 8,481 accidents—23 each and every day of the year. Few people would risk flying anymore and the industry would face shutdown.

At a loss of about one-third of everything flowing through it, the global food supply chain is a *two-sigma system*. Imagine every third piece of mail you sent was lost. Every third time you purchased a car it came without an engine. These are things that we would go out of our way to avoid. Indeed, it's difficult to find comparable examples of two-sigma systems because product waste and financial loss are so great that such systems simply don't exist for long. Nobody tolerates a two-sigma system.

Except that humankind owns one, and it's charged with delivery of a life-sustaining product: our food.

As financial returns begin to align with our better humanitarian and environmental impulses, the global food chain is a process ripe for rapid and dramatic improvement.

[1] Daniel Speckhard, "From Food Security to National Security," United Nations Committee for FAO, 2015, http://www.worldfooddayusa.org/daniel_speckhard.

[2] Xiaojun Wang and Kiyoshi Taniguchi, "Does Better Nutrition Cause Economic Growth?: The Efficiency Cost of Hunger Revisited," November 14, 2002, http://www.fao.org/docrep/007/ae030e/ae030e00.htm.

[3] "The Cost of Hunger in Africa: Social and Economic Impact of Child Undernutrition in Egypt, Ethiopia, Swaziland and Uganda," United Nations Economic and Social Council Economic Commission for Africa & Africa Union, February 26, 2014, http://www.uneca.org/sites/default/files/uploaded-documents/COM/com2014/com2014-the_cost_of_hunger-english.pdf.

[4] *2014-2015 Global Food Policy Report,* Washington, D.C.: International Food Policy Research Institute, 2015, 15.

[5] "Malnourished Economy: Global Hunger Leading to $2 Trillion Loss in World GDP, Albawaba Business, April 12, 2015, http://www.albawaba.com/business/malnourished-economy-global-hunger-leading-2-trillion-loss-world-gdp-681180.

[6] "2014 Global Nutrition Report: Actions and Accountability to Accelerate the World's Progress on Nutrition," The International Food Policy Research Institute, 2014, http://content.yudu.com/Library/A37tp2/GlobalNutritionRepor/resources/index.htm?referrerUrl=http%3A%2F%2Fwww.ifpri.org%2Fpublication%2Fglobal-nutrition-report-2014. See http://www.ifpri.org/pressrelease/return-investment-nutrition-high-so-why-do-so-many-governments-fail-adequately-invest-i.

[7] Donald S. Shepard et al, "Hunger in America: Suffering We All Pay For," Center for American Progress, October 2011, https://www.americanprogress.org/wp-content/uploads/issues/2011/10/pdf/hunger_paper.pdf.

[8] *Food Wastage Footprint: Full-cost Accounting,* Final Report, FAO, 2014, http://www.fao.org/3/a-i3991e.pdf, 7-8.

[9] A new estimate of 2.7 billion metric tons of CO_2 equivalent for GHG uses a different calculation method but is largely consistent with the 3.3 Gt commonly used. Likewise, water wastage is calculated at 300 km3 vs. the older 250 km3 because it includes some additional commodities and a different database. *Food Wastage Footprint: Full-cost Accounting,* Final Report, FAO, 2014, http://www.fao.org/3/a-i3991e.pdf, 10, 34.

[10] This is the annual return for the NYSE composite index. See NYSE Composite Stock Market Index Historical Graph, *ForecastChart.com,* http://www.forecast-chart.com/historical-nyse-composite.html.

[11] OECD is the Organisation for Economic Co-operation and Development, a forum of 34 countries describing themselves as committed to democracy and the market economy. It is primarily the developed nations of North America and Europe.

[12, 15, 16, 19] *Mitigation of Food Wastage: Societal Costs and Benefits,* FAO, 2014, http://www.fao.org/3/a-i3989e.pdf, 6, 11-15, 21-25, 28-29.

[13] Rick Fedrizzi, telephone interview with the authors, May 8, 2015.

[14] "World Green Building Trends SmartMarket Report: Business Imperative and Market Demand Driving Green Building Growth," Construction.com, February 28, 2013, http://construction.com/about-us/press/world-green-building-trends-smartmarket-report.asp.

[17] "Food Losses and Waste in the Context of Sustainable Food Systems," a report by the High Level Panel of Experts on Food Security and Nutrition of the Committee on World Food Security, Rome, 2014, http://www.fao.org/3/a-i3901e.pdf, 61.

[18] Nicolas Denis et al, "From Liability to Opportunity: How to Build Food Security and Nourish Growth," McKinsey & Co., March 24, 2015, http://www.mckinsey.com/insights/food_agriculture/from_liability_to_opportunity_how_to_build_food_security_and_nourish_growth.

CHAPTER 9: FOOD SECURITY IS NATIONAL SECURITY

"FOOD IS ABOUT TO DEMAND YOUR ATTENTION."

This was just one of several provocative observations made by author Michael Pollan in an open letter to President-elect Barack Obama in 2008. The era of cheap food might well be coming to an end, Pollan wrote, and with it a period when food is freely traded among nations. "Expect to hear the phrases 'food sovereignty' and 'food security' on the lips of every foreign leader you meet."[1]

Food has long been recognized as a building block of social and political well-being. The first-century Roman poet Juvenal understood that providing free wheat and lively entertainment—what he called "bread and circus"—was an effective way to build political power. In 18th-century France, a poor harvest led to higher food prices, which helped spark the French Revolution. In 1848 the most widespread revolutionary wave in European history was stirred by serious crop failures.

Food remains a powerful social and political force in modern times. When food prices reached historic highs in 2001, protests across the Middle East and Africa led to the ousting of leaders in Egypt and Tunisia. A more recent spike in global food prices became an important factor in the Arab Spring, a wave of uprisings that swept Arab countries in the Middle East and North Africa in late 2010.

WHEN A NATION IS UNABLE TO FEED ITSELF IT IS UNABLE TO PROTECT ITS PEOPLE

When a nation is unable to feed itself, it is unable to protect its people. It becomes subject to the whims of weather, commodity markets and the dictates of other governments. It becomes internally fragile. Reducing food waste enhances food security, which, in a very real sense, is national security.

THE POLITICS OF BREAD

When Michael Pollan wrote to the president-elect, the world was in the midst of one of the steepest food price increases in history. From January 2004 to May 2008 the price of rice increased 224 percent, wheat 108 percent and corn 89 percent.[2] This rocked the international community. New export restrictions in producer nations caused panic buying among major importers. Countries without the infrastructure to store their excess purchases saw massive food loss and waste.[3] Thirty-six countries appealed for food aid. Forty-eight countries experienced protests and riots.

The Middle East and North Africa are regions highly dependent on food imports. Egypt is one of the world's largest per capita consumers of bread, but grows only about 60 percent of what it consumes annually.[4] The Arab world in total must import about half the food that it requires each year. Fluctuations in food prices take an enormous toll on this region and contribute to an underlying sense of food insecurity. It's no surprise that the food price spike in 2007 and 2008 brought demonstrations and riots to Bahrain, Yemen, Jordan and Morocco.[5] Protesters gathering in Egypt's Tahrir Square clamored for "bread, freedom and social justice."[6]

A prolonged civil war in Syria is the result in part

of devastating droughts between 2006 and 2010 that ruined the livelihoods of half of the country's farmers and herders.[7] Crowds protested that the government took their "loaf of bread."[8]

The food price spikes of 2007 and 2008 triggered food riots in at least 14 African countries, especially those with higher poverty and greater urbanization. The link between hunger and poverty is widely recognized, but urbanization presents an entirely new wrinkle to national governments: As cities grow, populations become more concentrated and organized. This gives voice to more people and magnifies their message. One analysis suggests that Niger avoided riots in 2007 only because it lacked a large urban center.[9]

While food prices normally fluctuate by harvest and season, experts see little relief from continued high prices and supply disruption. After all, the global population is growing and demanding more and better calories. We're witnessing a Livestock Revolution, grains being diverted for biofuel, an ongoing competition for arable land and freshwater, and the rise of a global urban middle class. Signs also suggest more extreme weather as the result of climate change. One forecast has the world warming by three to four degrees by 2050. If this happens, sub-Saharan crop yields could decline by 15 to 20 percent, and in some countries dependent on rain-fed crops, by up to 50 percent as early as 2020.[10] Another study suggests that reduced agriculture means reduced employment and an increased pool of potential combatants if conflict arises.[11]

Even in locations with stable governments, modern agriculture and strong social safety nets, food shortages and high prices place enormous pressures on citizens and leaders.

An economist might explain the issue by saying that the demand for food is relatively inelastic. What this means in human terms is that we all have to eat, even if the price of food skyrockets. In places with high food security like Houston, Texas, poor residents cope by trading quality nutritious calories for cheap ones, visiting food banks or skipping meals. In more fragile regions, a spike in food prices can plunge millions of people into poverty over-night. As always, it seems the people who are hurt most are those already least able to cope. Studies in Bangladesh and Malawi indicate that smallholder farmers lose more than large landholders when the price of staple crops increases.[12] The tie between conflict and food is also direct and tragic: Children in Zimbabwe and Burundi who experienced violent conflict were substantially more apt to be stunted. Even relatively short-term violence can disrupt the food chain and scar a child for life.[13]

A MORE RESILIENT FOOD SUPPLY SYSTEM

Unfortunately, the world remains a very unsettled place. Leaving aside weather extremes and natural disasters, humankind has countless ways of inflicting harm upon itself. The "Conflict Barometer" of the Heidelberg Institute for International Conflict Research reported for 2014 that there were 424 global conflicts, 223 of them violent.[14] Twenty-five were classified as limited wars and 21 as wars. Almost a quarter of all conflicts were primarily over natural resources and raw materials. Another 50 were over territory. The competition for land and water, along with widespread food insecurity, is at the very heart of poverty and violence.

The world has developed powerful mechanisms for coping with disaster. International humanitarian relief grew to a record $22 billion in 2013. Governments accounted for $16.4 billion of this and private sources the remainder. Some three-quarters went to long-term assistance, almost entirely in countries with high levels of poverty and food insecurity. In fact, the majority of assistance went to provide basic goods and services, with food aid ranging between one-quarter and one-third of spending in any given year.[15]

This $22 billion in humanitarian aid is almost surely understated. In some countries relief is largely internal: From 2009 to 2012, India spent $7 billion for disaster relief and risk reduction, only $137 million coming from international humanitarian agencies.[16]

Yet even with these extraordinary spending levels, over one-third of humanitarian needs remain unmet. People continue to suffer from the tightly interwoven threats of conflict, political instability, lack of land and freshwater, and insufficient food.

Humankind requires new strategies. The U.S. State Department concluded after the food price crisis of 2007 and 2008 that public and private investment was required to improve agricultural productivity and meet the rising demand for food. Spending initiatives might focus on new agricultural technologies and on finding ways to break bottlenecks and enhance transportation and distribution along the food supply chain.[17] Said differently, the food supply chain needs to be more *resilient*. Weather and the economy may be uncooperative, but our ability to produce, store and distribute food can become significantly more robust. The food supply chain can become *elastic,* whereas our hunger cannot.

Initiatives like those proposed by the U.S. State Department will inevitably reduce loss and waste. We can reclaim up to one-third of all food produced, enhancing food security and national security. This would be investment spending that precedes, and can help offset the need for, humanitarian aid. As officials noted, "The U.S. and other donors spend much more money responding to humanitarian disaster than in investing in building more resilient communities."[18]

China is an example of a country making investments designed to create a more resilient food supply chain. The country faces the herculean task of feeding 22 percent of the world's population with just 9 percent of the earth's arable land,[19] a resource declining by as much as 3.7 million hectares each year.[20] China's annual harvest of 115 million tons of wheat is the largest in the world, used almost exclusively to feed its own population. Extreme weather is a constant threat to food security.

China's response over time has been to increase its grain storage capacity. The country now stockpiles as much as 85 million tons, or two-thirds of its annual wheat demand. This provides a buffer against weather and price shocks[21] and enhances China's national security.

We know from earlier chapters that cereal loss and waste in Industrial Asia and South and Southeast Asia rank second and third, respectively, in total lost food, behind only vegetable waste in Industrial Asia.[22] Because of their enormous volumes, wasted cereals also rank first in contributing to the global carbon footprint.[23]

If China loses and wastes up to one-third of its annual cereal crop before it can be consumed,[24] this translates to 38 million tons of wheat. Improving infrastructure to reduce wheat loss to a negligible amount would translate to an extra annual harvest every fourth year, or refilling the country's enormous stockpiles in just over two years. This is without another hectare being planted or drop of water being consumed by agriculture. The enhancements to yield, carbon footprint and food security would be dramatic. These translate directly to strengthened national security.

China has expanded its strategic food reserve system to include crops like rice, and its program now extends across 31 provinces.[25] In addition, the country has opened up supply sources for grain in Brazil, Argentina and Africa.[26]

Dependence on imports by Arab nations is expected to increase by almost 64 percent by 2030.[27] Nations in the region are taking important steps to build a more resilient food system. The United Arab Emirates now stores 12 weeks of wheat, rice and powdered milk. The government also requires private retailers to stock a month's worth of poultry and fresh vegetables to meet short-term shocks.[28] Some countries have extended their social safety nets. Others are educating families to expand their diets which, on average, derive 35 percent of their calories from wheat. Investments are also being made to improve cereal yields, which currently average only half of the worldwide rate. Steps are also being taken to improve the production, storage, transport and marketing of food, which constitutes about 75 percent of its retail price.[29] Some countries are using financial instruments to hedge commodity prices and reduce stockpiling. And like China, some nations are acquiring land around the world: Saudi Arabia and the United Arab Emirates hold more than 2.8 million hectares in Indonesia, Pakistan and the Sudan.[30]

ENHANCING NATIONAL SECURITY

Food is the largest sector of the world economy. When the global food system is jolted, humankind suffers. People panic. Governments respond to protect their citizens. The connection between hunger, waste and conflict is stark and indisputable. Wasted food means wasted water: Countries that share river basins are more likely to go to war.[31] Wasted food means squandered land: Wars are waged regularly over the scarcity of arable land. Sixty-five percent of the world's food-insecure people live in just seven countries: India, China, Bangladesh, Indonesia, Pakistan, Ethiopia and the Democratic Republic of the Congo. Experts link food insecurity to increased risk of democratic failure, protest and rioting, communal violence and civil conflict. A study by the World Food Programme concludes that "violent conflicts, in turn, create food insecurity, malnutrition and—in some instance—famine. Thus food insecurity can perpetuate conflict."[32]

The leading light of the Green Revolution, Norman Borlaug, once said, "If you desire peace, cultivate justice, but at the same time cultivate the fields to produce more bread; otherwise there will be no peace."[33] As we shape the next food revolution, we might add, "And if you desire more bread, create a green, sustainable and robust food system that preserves and protects what you work so hard to produce." When we consider reducing food waste as an element of national security, we value all the stakeholders, from national governments and urban populations to smallholder farmers and children wasting away from lack of nutrients. We value the land, freshwater and the environment.

We learned earlier that in Afghanistan, simple metal silos for grain storage produced by local tinsmiths reduced food losses from as much as 20 percent to 1-2 percent.[34]

What would just a quarter of annual humanitarian aid, say $5 billion, do to strengthen the global food system and enhance food security? What would spending prior to suffering do to reduce the need to relieve suffering? What would this mean for peace between nations and a stronger international community?

The weather may become more extreme. The global population may exceed 9 billion. The demands of hunger may remain inelastic. But our willingness to build a robust, flexible and elastic food supply chain will reduce food waste—and reducing food waste will enhance global food security. As many governments learned after the 2008 food crisis, food security truly is national security.

[1]Michael Pollan, "Farmer in Chief," *The New York Times,* October 12, 2008, http://michaelpollan.com/articles-archive/farmer-in-chief/.

[2, 3]"Food Prices Crisis of 2007-2008: Lessons Learned," U.S. Department of State, Washington, D.C., March 3, 2011, http://www.state.gov/r/pa/prs/ps/2011/03/157629.htm.

[4]Julia Berazneva and David R. Lee, "Explaining the African Food Riots of 2007-2008: An Empirical Analysis," Charles H. Dyson School of Applied Economics and Management," Cornell University, March 2011, http://www.csae.ox.ac.uk/conferences/2011-edia/papers/711-berazneva.pdf.

[5]"Let Them Eat Baklava," *The Economist,* May 17, 2012, http://www.economist.com/node/21550328.

[6]"Where's the 'Bread, Freedom and Social Justice' A Year After Egypt's Revolution?", The Guardian, January 25, 2012, http://www.theguardian.com/global-development/poverty-matters/2012/jan/25/egypt-bread-freedom-social-justice.

[7, 10, 12]*2014-2015 Global Food Policy Report,* Washington, D.C.: International Food Policy Research Institute, 2015, http://www.ifpri.org/sites/default/files/publications/gfpr20142015.pdf, 6, 28.

[8]Toni Verstandig, "Food Security is National Security," *The World Post,* June 24, 2014, http://www.huffingtonpost.com/toni-verstandig/food-security-is-national_b_5540015.html.

[9]Julia Berazneva and David R. Lee, "Explaining the African Food Riots of 2007-2008: An Empirical Analysis," Charles H. Dyson School of Applied Economics and Management," Cornell University, March 2011, http://www.csae.ox.ac.uk/conferences/2011-edia/papers/711-berazneva.pdf.

[11, 13]Henk-Jan Brinkman and Cullen S. Hendrix, "*Food Insecurity and Violent Conflict: Causes, Consequences, and Addressing the Challenges,*" World Food Programmed Occasional Paper No. 24, July 2011, 6, 12.

[14]*Conflict Barometer 2014,* Heidelberg Institute for International Conflict Research, 2015, http://www.hiik.de/en/konfliktbarometer/pdf/ConflictBarometer_2014.pdf.

[15, 16]*Global Humanitarian Assistance Report 2014,* devinit.org, accessed 2015, http://www.globalhumanitarianassistance.org/report/gha-report-2014, 70, 7.

[17]"Food Prices Crisis of 2007-2008: Lessons Learned," U.S. Department of State, Washington, D.C., March 3, 2011, http://www.state.gov/r/pa/prs/ps/2011/03/157629.htm.

[18]Jack Bobo and Chris Hegadorn, "Food Security is National Security," *The Plate-National Geographic,* October 21, 2014, http://theplate.nationalgeographic.com/2014/10/21/food-security-is-national-security/.

[19, 21]Jim Harkness, "Food Security and National Security: The Rest of the World Should Learn From China's Approach to Managing Wheat Supplies," February 23, 2011, http://www.iatp.org/documents/food-security-and-national-security-learning-from-china%E2%80%99s-approach-to-managing-its-wheat-s.

[20, 26]Jacques Carles and Paul-Florent Montfort, "Food Security and National Defense, A Geopolitical Perspective," Momagri, accessed 2015, http://www.momagri.org/UK/points-of-view/Food-Security-and-National-Defense-A-Geopolitical-Perspective_661.html.

[22, 23]*Food Wastage Footprint: Impacts of Natural Resources, Technical Report* (Working Document), FAO, http://www.fao.org/3/a-ar429e.pdf, 15, 18.

[24]"Agricultural Official Condemns Unnecessary Loss and Waste of Good Food," Food and Agricultural Organization of the United Nations, April 3, 2014, http://www.fao.org/news/story/en/item/219025/icode/.

[25, 28]Nicolas Denis et al, "From Liability to Opportunity: How to Build Food Security and Nourish Growth," McKinsey & Co., March 24, 2015, http://www.mckinsey.com/insights/food_agriculture/from_liability_to_opportunity_how_to_build_food_security_and_nourish_growth.

[27, 29, 30]*Improving Food Security in Arab Countries,* Washington, D.C.: The World Bank, 2009, "Executive Summary," http://siteresources.worldbank.org/INTMENA/Resources/FoodSecfinal.pdf.

[31, 32]Henk-Jan Brinkman and Cullen S. Hendrix, "*Food Insecurity and Violent Conflict: Causes, Consequences, and Addressing the Challenges,*" World Food Programme Occasional Paper No. 24, July 2011, http://ucanr.edu/blogs/food2025/blogfiles/14415.pdf, 6, 20.

[33]*Brainy Quote,* Norman Borlaug, http://www.brainyquote.com/quotes/quotes/n/normanborl372057.html#kPfgmtdiWQa5J7IG.99.

[34]Tadele Tafera, "The Metal Silo: An effective grain storage technology for reducing post-harvest insect and pathogen losses in maize while improving smallholder farmers' food security in developing countries," The International Maize and Wheat Improvement Center, accessed 2015, http://www.researchgate.net/publication/229293163_The_metal_silo_An_effective_grain_storage_technology_for_reducing_post-harvest_insect_and_pathogen_losses_in_maize_while_improving_smallholder_farmers_food_security_in_developing_countries, and Brian Lipinski et al, "Reducing Food Loss and Waste, World Resources Institute, working paper, June 2013, http://www.wri.org/publication/reducing-food-loss-and-waste, 101.

CHAPTER 10: REDUCING FOOD WASTE: THE LOW-HANGING FRUIT

HUMANKIND CONSUMES 7.5 TRILLION CALORIES OF FOOD A YEAR.[1]

That's an average of 2,860 kcal/person/day, more than enough to meet the minimum dietary energy requirements set by the United Nations FAO.[2] Those figures don't include any of the 2 billion metric tons of food grown for animal feed, industry use and biofuels. And they obviously exclude the one-third loss and waste that we know leaks out of the food supply chain each year.

Despite those enormous exclusions, everyone on Earth today—in theory and on average—has more than enough food to eat.

By 2050 we expect 9.6 billion people to be living on Earth.[3] If we simply freeze the global agricultural footprint at today's output but reduce by 50 percent the loss and waste we currently suffer, humankind would have 9.1 trillion calories[4] of food available. That's enough to provide every person on earth in 2050 with 2,594 kcal/person/day, still in excess of the FAO's minimum requirements.

Of course, we can't feed people by spreadsheet calculations. Statistics and abstractions don't work. Just like land and freshwater, the big picture can obscure a series of smaller, more complicated truths. The fact is that over 800 million people are chronically hungry while between 2 billion and 3 billion suffer from micronutrient deficiencies. In Kenya the average food deficit is 156 kcal/person/day, and in Pakistan it is 169 kcal/person/day. In North Korea the deficit is 303 kcal/person/day.[5]

However, even in the abstract, we should hold on to this truth: *Our global food system is capable of feeding everyone in the world today and tomorrow.* By reducing loss and waste, it has fantastic expansion possibilities

OUR GLOBAL FOOD SYSTEM IS CAPABLE OF FEEDING EVERYONE IN THE WORLD TODAY AND TOMORROW

without increasing our agricultural footprint. A more robust food supply chain would also support the growing shift from carbohydrates to proteins and fats, ensuring nutritious diets. In addition, the food grown today for animals and industry could potentially feed another 4 billion people.

This is the fundamental dilemma that leaders around the globe recognize: There is food for everyone, yet not everyone can eat. Pope Francis called the world's attention to this issue by citing the "paradox of abundance,"[6] the eloquent words of his predecessor Pope John Paul II that capture so well today's problem.

This also suggests that we need to challenge the pervasive assumption that agriculture must expand by 60 percent or more to feed the world's population in 2050.[7] Given what we know about food loss and waste, this is like working harder to shovel sand into a barrel—a barrel with a hole in its bottom.

The question is, how do we close the gap between an abstract, "average" spreadsheet world with enough calories for everyone and a real world where every human being would have genuine food security?

We already know the crucial first step.

CLOSING THE GAP: FOOD MUST BE AFFORDABLE

"People who are hungry are poor; people who are poor are hungry." That's the poignant reminder from Catherine Bertini, former executive director of the United Nations World Food Programme. "Hunger and poverty are totally interchangeable," she says. "And the longer one generation stays that way the more likely the next generation will be the same. There isn't any more important driver of hunger than poverty."

Bertini's agency focused on some of the poorest, most food-insecure countries and regions on Earth. But hunger and poverty are inextricably linked even in the prosperous city of Houston, Texas. Brian Greene of the Houston Food Bank feeds people every day who must sacrifice their food money to pay rent or keep their utilities running.

When people are poor, they are inevitably hungry. If we want to feed the world, food must first be affordable.

It's hard to imagine a more complex, stubborn problem than global poverty. We have seen already that solutions to it must be holistic and integrated, like the work being done by Babban Gona to create new economic models for Nigerian farmers. Solutions must be customized to each people and culture, like the Grameen Bank established by Muhammad Yunus to provide credit to farmers in Bangladesh and create social entrepreneurs. "Poverty does not belong in a civilized human society," Yunus says. "The real issue is creating a level playing field for everybody—rich and poor countries, powerful and small enterprises—giving every human being a fair chance."[8]

That means effective solutions require global cooperation, public-private partnerships, resolute political and social will, peace and security. Today's successful programs include initiatives to support rural development, provide social protection for the most vulnerable members of communities, foster education, strengthen resilience to conflict and natural disasters, promote the health of mothers and children, build infrastructure, improve agricultural inputs, restore land, provide access to freshwater, and introduce new technologies and market opportunities. Finding ways to support smallholder farmers has been a particularly fruitful strategy in the last generation for reducing both poverty and hunger.

There is no question that we are making progress and have momentum. Between the early 1980s and 2005, a half-billion people rose above the poverty level.[9] Since then, some 70 million people escape poverty each year. China in particular has made remarkable progress. Other regions of the world are also forging ahead: In 2013 at the first summit of the Community of Latin American and Caribbean States, heads of state developed a plan to eradicate poverty and endorsed a zero hunger target for 2025. In July 2014 at the African Union summit, African heads of state likewise committed to end hunger on the continent by 2025.[10] FAO Director-General José Graziano da Silva has endorsed inclusive food security strategies along with strengthened social protections, recognizing that increasing food production alone is not enough to end hunger.[11]

All of these initiatives help reduce poverty and make food affordable. But there's more we must do if we really wish to feed the world.

CLOSING THE GAP: FOOD MUST BE AVAILABLE

Compared with solving poverty, enhancing the global cold chain looks like a walk in the park.

True, it's not always easy to assemble the pieces required to seamlessly ship perishable foods. It's still impossible to build a modern cold chain in regions without good roads or electricity. But as a targeted and proven investment, it's hard to find a better way to reduce loss and waste, expand distribution and make food available to hungry people.

The cold chain also has a secret weapon: It specializes in the kind of foods that are essential to human health.

"If you simply asked 'how do you reduce the number of hungry people in the world' you'd say 'grow more rice because it has energy and it's already part of people's diets,'" says Dr. Charles Winkworth-Smith, co-author of a study by the University of Nottingham that focused on perishable food loss. "But that ignores micronutrient deficiencies which cause a huge problem, both economically and in human suffer-

HIDDEN HUNGER CAN COST A COUNTRY 1% OF GDP

ing. Hidden hunger can cost a country 1 percent of GDP," Winkworth-Smith adds. "But the real value of reducing waste in fruits and vegetables is to improve the nutritional value of diets to fight anemia, visual impairment or blindness and other crippling problems."[12]

This poses an interesting idea: What if some of the lowest-hanging fruit in solving world hunger and malnutrition turned out to be fruit? And vegetables. While they don't provide the caloric impact of cereals, fruits and vegetables address the micronutrient deficiencies that create hidden hunger in an estimated 2 billion people.

The University of Nottingham report highlights a number of troubling issues. In Pakistan, vitamin A status has deteriorated in recent years and there has been little or no improvement in other areas linked to micronutrient deficiencies. Three-quarters of children under 5—and virtually all infants under 6 months old—in Kenya suffer from anemia. Similarly, iron deficiency is a serious problem in India and Kenya, as is zinc deficiency in China. Worldwide, micronutrient deficiencies leave 1.2 billion people with weakened immune systems, 1.8 billion susceptible to reduced mental capacity, and over 200 million young children and pregnant women at risk of visual impairment or blindness.[13]

Winkworth-Smith and his team investigated the potential micronutrients that could be made available if more fruits and vegetables currently grown made it from farm to fork. They chose as a reasonable target saving 25 percent of the current loss and waste in just four countries: China, India, Pakistan and Kenya.

The results were extraordinary. As much as 53 million metric tons of food—about half of the total consumption of fruits and vegetables in Europe[14]—could be saved. This would provide complete caloric energy for up to 22 million people.[15] But more importantly, these savings would provide iron for up to 66 million people and vitamin A for as many as 70 million people.[16] Protecting fruits and vegetables also has important benefits to the environment;

spoiling vegetables in particular give up almost twice their volume in carbon emissions.[17]

"The figures that we've produced are pretty abstract," Winkworth-Smith says. "But the work we've done highlights the importance of fruits and vegetables to a healthy diet. They can really make a big difference."

This reinforces the cold chain's role in helping to solve the scourge of hidden hunger. Vegetables and fruits account for 40 percent of all food waste. Combined with meat, fish, seafood and dairy, the total is more than 50 percent. In addition, the shelf life of many starchy roots—staple foods in many developing nations—is enhanced with refrigeration. A cold chain created to protect fruits and vegetables can be deployed to preserve foods that comprise nearly three-quarters of the entire food chain—the most nutritious foods on the planet.[18]

And, there is another hidden but powerful factor that emerges when we focus on perishable foods that provide people with their essential nutritional needs. "Most of the greenhouse gas emission calculations for foodstuffs make comparisons like a pound of milk versus a pound of sugar," says Alexander J. Travis, faculty director for the environment, Cornell University's Atkinson Center for a Sustainable Future at Cornell University. "A pound of milk takes a lot more carbon and water to produce. But what's the relative nutritional value of each? If you look at nutritional value, all of a sudden dairy, eggs, poultry, fish—if produced sustainably and eaten in reasonable amounts—these things can actually be much more sustainable than the way we use a lot of our cropland."

These also happen to be high-nutrient and perishable

foods—just like fruits and vegetables—which benefit from a robust cold chain. "And so," Dr. Travis adds, "judgments about which foods are sustainable based on greenhouse gas calculations can flip when you start looking at it through the lens of nutrition. And from an environmental perspective we're not at all being smart about this right now. Where we produce certain foods, how we produce those foods, and how we get them to consumers are all areas where we need to change the way we think about sustainability."

His colleague Lauren Chambliss, communications director of the Atkinson Center, agrees and adds, "Just getting people to understand that by some estimates, we already produce enough food to feed the world and that nutrition, distribution and access to a healthy mix of foods are intimately interlinked, will be critical to human prosperity in 2050, not just in developing countries but at home as well."[19]

The modern cold chain has the added advantage of supporting our first goal, making food more affordable. Post-harvest loss in Africa is estimated to reduce the income of 470 million smallholder farmers by a least 15 percent.[20] The cold chain's growth in India could support banana exports and enhance the wealth of 34,600 smallholder farmers there.[21] By wasting less, the modern cold chain generates wealth as it reduces hunger and malnutrition.

The arithmetic of hunger seems abundantly clear: Simply producing enough calories is necessary but insufficient to adequately feed humankind. Only by making food—and the right kinds of nutritious food—both afford-able and available can we close our "spreadsheet gap" in feeding the world. And when all the arithmetic is done, it may well be that focusing on nutrition is the best way to measure the true sustainability of food production.

CLOSING THE GAP: FOOD MUST BE PRECIOUS

A study conducted toward the end of the Great Depression in 1939 showed that homes in Britain wasted less than 3 percent of their weekly groceries. A recent study showed food waste in United Kingdom households to be 25 percent. Research in the U.S. shows similar results.[22]

What's going on?

When food is inexpensive compared with disposable income, it's more apt to be wasted. Studies show time and again that as income rises, so too does the amount of food we discard.[23] And thanks to modern food retailing, consumers in developed countries have lofty expectations for what they find in their local grocery stores. Perfect tomatoes are advertised and perfect tomatoes appear in the store. When some are purchased, others quickly appear so that shelves are always fully stocked. Everything is affordable, and if the price does rise for one item, there are usually other nutritious foods to substitute. After a while, all that wonderful food so easily acquired becomes less precious and is taken for granted.

There's another related phenomenon going on in restaurants, especially in the U.S. Inexpensive, plentiful food—and human nature—have caused portion sizes to grow. Servings are often beyond what the average person can comfortably eat and sometimes beyond what is healthy. Twenty years ago bagels were 3 inches across and had 200 calories; today they can be 6 inches across and deliver 500 calories. Hamburgers are 23 percent larger and Mexican entrées 27 percent larger.[24] Restaurateurs do this because competition is stiff, American consumers equate quantity with value, and larger portions mean more customers.[25]

Not surprisingly, Americans today spend about one-fourth the income on food that they did in 1960.[26] And many misunderstand the amount of food they need to feel full and to be healthy.

When food is no longer precious, consumers buy, prepare and eat too much of it. They also fail to use it before it spoils.[27] These wasteful activities are a function of habit, something that responds to education. Retailers, restaurateurs and national governments around the world have taken up the cause.

In 2012 the European Parliament adopted a resolution to cut food waste in half by 2020 and designated 2014 as the "European year against food waste."[28] Individual nations have launched programs designed to complement and encourage this initiative. *Damn Food Waste* is a Netherlands campaign that encourages everyone from consumers and food service providers to farmers and politicians to prepare meals from food that would otherwise be wasted.[29] The British *Love Food, Hate Waste* campaign, created in conjunction with leading food retailers and brands, spent 15 million euros over five years, resulting in 500 euros of food waste avoided for every euro spent.[30] The third largest supermarket chain in France, Intermarché, launched a campaign in 2014 called "Inglorious Fruits and Vegetables" that celebrated the "ugly" produce often discarded by growers as unfit for consumption.[31] With discounts as high as 30 percent, the initial selection sold out immediately and increased store traffic. Likewise, Canada's largest food retailer, Loblaws, launched its "Naturally Imperfect" product line by offering 30 percent discounts on ugly apples and potatoes, with plans to expand to other produce varieties.[32]

In Great Britain, Sainsbury's has fashioned an entire consumer initiative around what it calls "the rise of new-fashioned values." In surveying and talking with its 22 million customers, the retailer is finding that shoppers now focus on "savvy sustainability." They rate "reducing waste" high in what they value in a food retailer. Ninety percent create a shopping list before they visit the store, and four out of six plan meals for the full week. Meanwhile, Sainsbury's is encouraging its shoppers to "love your left-overs" and, like Intermarché, to buy and enjoy "ugly" fruits and vegetables. The retailer has changed its policies and now buys all British fruits and vegetables that meet regulations and standards, even if they're not cosmetically perfect.[33]

Sainsbury's is also taking steps within its 300 stores to put all waste to positive use. The retailer donates all surplus food fit for human consumption to charities— the equivalent of more than 1.2 million meals. All unsold bread and bakery products are turned into animal feed for the farming industry. And the company's Cannock store in Staffordshire became the first U.K. retail outlet to be powered by food waste alone; it uses electricity generated from anaerobic digestion, a process that turns food waste into bio-methane gas. Since 2013 Sainsbury's has sent zero waste to landfills from its stores, depots and store support centers.[34]

Legislation in the United States is also creating positive change. A law in Massachusetts prohibits any institution producing more than 1 ton of food waste per week from sending it to a landfill. Instead, edible food must be donated and the rest shipped to anaerobic digestion facilities or distributed to farmers to use as animal feed.[35] In New York state, a program offers incentives of up to $2 million per installation for farmers to install anaerobic digestive equipment to convert organic waste produced by cows into electricity.[36] On a national level, the U.S. Department of Agriculture (USDA) and the U.S. Environmental Protection Agency launched the U.S. Food Waste Challenge in June 2013, which now has over 1,000 participants. In addition, the USDA has set goals to minimize food waste in school meals programs, educate consumers on ways to reduce waste, recover food removed from commerce (such as supporting donations of whole-some but misbranded meat and poultry products), and conduct ongoing research. Much of this new research is directed at finding new ways to reduce perishable product waste.[37]

The world's largest food retailer, Walmart, has also taken a leadership role in reducing food waste at the retail and consumer levels. The company donated 571 million pounds of food to food banks and hunger relief efforts in 2013.[38] Most recently, Walmart and the Walmart

Foundation committed to provide nutrition education to 4 million U.S. households.[39] Other industry initiatives involve providing advice on food storage and clarifying food date labeling. It has become apparent that in many cases consumers simply do not understand how much food is being wasted. Once educated, they are ready and willing to adjust their buying and food preparation habits.

As a general rule, global food retailers are seeking first to reduce food waste, second to reuse, and third to recycle and recover. Only when all those fail are landfills deemed an option.

U.S. restaurants and their patrons are beginning to make constructive changes as well. In the National Restaurant Association's "What's Hot in 2014," half-portions and smaller portions were identified as a top trend.[40] More restaurants are offering dishes as light as 400 calories, and some are helping their patrons control portions through simple changes like smaller bowls and plates. Many customers have begun to feel comfortable sharing meals.

Launched in 2012, Menus of Change™ is a national educational project developed by The Culinary Institute of America and Harvard School of Public Health, Department of Nutrition. Seeking collaboration across the entire food service industry, the program has gathered together executives, chefs, nutrition and environmental scientists, farm and fisheries experts, and policymakers. Menus of Change principles are holistic in nature, anticipating that "fast-moving, mid- and long-term global trends—from continued population growth and increasing resource shortages to commodity price spikes and food security issues—will increasingly reframe how we think about food and food service in the United States."[41] Specific strategies include reducing portion sizes, increasing offerings of fruits and vegetables, providing transparency in food sources and nutritional value, rewarding best practice agricultural production, and promoting health and sustainability through menu choices.

Even though France is known for its culinary excellence, food waste is still a problem that costs the country some $21 billion annually. A recent report prepared for the French government included 36 proposals aimed at changing the "almost automatic" habit of restaurants throwing away leftovers.[42] One especially striking proposal called for "le doggy bag." This proposal is anathema in a country that has long favored quality over quantity. However, "le doggy bag"—or what some in France hope will be called "le gourmet bag"—is another indication that governments in developed nations have taken the problem of food waste seriously. Teaching consumers new habits that encourage them to value the affordable food all around them is one effective solution.

FROM SPREADSHEET TO REALITY

Food journalist Mark Bittman described the current food system as being "geared to letting the half of the planet with money eat well while everyone else scrambles to eat as cheaply as possible."[43] This is one reason that our "average" spreadsheet calculations are so rich in calories but fall short of reality. Closing this gap—this "paradox of abundance"—requires three steps:

- Reducing poverty to make food affordable.
- Reducing loss and waste to make food available to more people and, with an expanded cold chain, allow for vast improvements to global nutrition.
- Helping consumers recognize food as a precious resource to begin to repair one of the most broken segments of the food supply chain.

The result of these strategies is that more people can be fed, nutrition will improve, and there will be far less stress on our agricultural resources, urban landfills and global environment. By using our current system more efficiently, we can feed the world's population today and tomorrow. This is the lowest-hanging fruit when it comes to solving the puzzle of food loss and waste.

[1] 7.2 billion people x 365 days/year x 2,860 kcal/person/day consumed in 2015. See Nikos Alexandratos and Jelle Bruinsma, "World Agriculture Towards 2030/2050 (The 2012 Revision)," ESA Working Paper No. 12-03, Agricultural Development Economics Division, Food and Agriculture Organization of the United Nations, June 2012, http://www.fao.org/docrep/016/ap106e/ap106e.pdf, 23.

[2] "The State of Food Insecurity in the World: Questions and Answers," FAO, Rome, 2014, http://www.fao.org/publications/sofi/2014/2014faqs/en/. "The Minimum Dietary Energy Requirement (MDER) is a country-specific threshold that FAO employs as a cut-off point to estimate the prevalence of undernourishment. It is specific for age and sex. It includes the energy expended by the human body in a state of rest and adding in a factor to account for physical activity. In the 2006-08, no country's MDER exceeded 1,980 kcal/person/day. A recent spreadsheet can be found at http://www.fao.org/search/en/?cx=018170620143701104933%3Aqq82jsfba7w&q=mder&cof=FORID%3A9.

[3] "World Population Projected to Reach 9.6 Billion by 2050," United Nations Department of Economic and Social Affairs, June 13, 2013, http://www.un.org/en/development/desa/news/population/un-report-world-population-projected-to-reach-9-6-billion-by-2050.html.

[4] 2,860 kcal/person/day / (.66 + (.33 x .5) = 3,467 kcal/person/day by recovering 50 percent loss and waste. 3,467 x 365 days/year x 7.2 billion people in 2015 = 9.11 trillion calories that could be consumed today if loss and waste were reduced by 50 percent.

[5] "Depth of the food deficit (kilocalories per person per day)," The World Bank, accessed 2015, http://data.worldbank.org/indicator/SN.ITK.DFCT.

[6] "Pope Focuses on Paradox of Abundance in Address to Experts," Las Vegas Sun, February 7, 2015, http://lasvegassun.com/news/2015/feb/07/pope-focuses-on-paradox-of-abundance-in-address-to/.

[7] "Food Wastage Footprint: Impacts on Natural Resources," FAO, 2013, http://www.fao.org/docrep/018/i3347e/i3347e.pdf.

[8] Muhammed Yunus, Banker To The Poor: Micro-Lending and the Battle Against World Poverty, Public Affairs, 2007, 248.

[9] Laurence Chandy and Geoffrey Gertz, "With Little Notice, Globalization Reduced Poverty," Yale Global Online, July 5, 2011, http://yaleglobal.yale.edu/content/little-notice-globalization-reduced-poverty.

[10, 11] "Strengthening and Enabling Environment for Food Security and Nutrition," The State of Food Insecurity in the World 2014," FAO, IFAD and WFP, Rome, 2014, http://www.fao.org/3/a-i4037e.pdf.

[12] Dr. Charles Winkworth-Smith, telephone interview with the authors, April 15, 2015.

[13, 14, 16, 21] C.G. Winkworth-Smith et al, "The Potential Value of Reducing Global Food Loss," The University of Nottingham, Division of Food Sciences, School of Biosciences, March 2015, 26-27, 29, 30.

[15]Assumptions in the report include 1) all food is eaten raw, and 2) the "average person" is an adult male 19-30 years old who requires 2,100 kcal/day.

[17, 18]*Food Wastage Footprint: Impacts of Natural Resources, Technical Report* (Working Document), FAO, http://www.fao.org/3/a-ar429e.pdf, 103, 117, 20.

[19]Lauren Chambliss, interview with the authors, May 12, 2015

[20]Reducing Global Food Waste and Spoilage, The Rockefeller Foundation and Global Knowledge Initiative, May 2014, http://postharvest.org/Rockefeller%20Foundation%20Food%20Waste%20and%20Spoilage%20initiative%20Resource%20Assessment_GKI.pdf.

[22, 23, 27]Julian Parfitt et al, "Food Waste Within Food Supply Chains: Quantification and Potential for Change to 2050," Philosophical Transactions of The Royal Society, September 27, 2010, http://rstb.royalsocietypublishing.org/content/365/1554/3065.full, 3074, 3077.

[24]Wendy Rotelli, "Why Are American Food Portions So Big?" Restaurants.com, March 27, 2013, https://www.restaurants.com/blog/why-are-american-food-portions-so-big/#.VTFFbvnF_cM.

[25]"Why Are Portions in U.S. Restaurants So Big? Ask USA TODAY," June 2, 2014, http://www.usatoday.com/story/opinion/2014/06/02/food-portions-restaurants/9734413/.

[26]Heidi Stevens, "A Problem of Grand Proportions," *Chicago Tribune,* November 2, 2011, http://www.usatoday.com/story/opinion/2014/06/02/food-portions-restaurants/9734413/.

[28]Dana Gunders, "Wasted: How America is Losing Up to 40 Percent of Its Food from Farm to Fork to Landfill," NDRC Issue Paper, August 2012, http://www.nrdc.org/food/files/wasted-food-ip.pdf.

[29, 30]Joris Tielens and Jeroen Candel, "Reducing Food Wastage, Improving Food Security?", Food & Business Knowledge Platform, July 2014, http://knowledge4food.net/wp-content/uploads/2014/07/140702_fbkp_report-foodwastage_DEF.pdf.

[31]Martha Cliff, "Forget the Ugly Fruit, Meet the Ugly Fruit Bowl!: French Supermarket Introduces Lumpy and Misshapen Fruit and Vegetables—Sold At a 30% Discount—To Combat Food Waste," July 16, 2014, http://www.dailymail.co.uk/femail/food/article-2693000/Forget-ugli-fruit-meet-ugly-fruit-bowl-French-supermarket-introduces-lumpy-misshapen-fruit-vegetables-sold-30-discount-combat-food-waste.html#ixzz3DbAhw2fm.

[32]Caitlin Troutt, "The Ugly Truth: Imperfect Produce Tastes Great and This Retailer is Selling it at a Discount," FoodTank.com, April 27, 2015, http://foodtank.com/news/2015/04/canadian-retailer-begins-selling-ugly-produce-at-a-discount.

[33]"The Rise of New-Fashioned Values," Sainsbury's, accessed 2015, http://www.j-sainsbury.co.uk/media/1488636/csr_factsheet_new_fashioned_values.pdf.

[34]"Sainsbury's 20x20 Factsheet Quarter 3 2014/15, http://www.j-sainsbury.co.uk/media/2386154/factsheet_environment.pdf.

[35]Hannah Ritchie, "Massachusetts Imposes Ambitious Law to Eliminate Commercial Food Waste," *Sustainable Brands,* August 11, 2014, http://www.sustainablebrands.com/news_and_views/waste_not/hannah_ritchie/massachusetts_imposes_ambitious_law_eliminate_commercial_foo.

[36]"Governor Cuomo Announces Initiatives to Help New York Dairy Farmers Increase Profitability and Reduce Energy Costs," New York State, February 1, 2013, https://www.governor.ny.gov/news/governor-cuomo-announces-initiatives-help-new-york-dairy-farmers-increase-profitability-and.

[37]"USDA's Commitments and Deliverables Through 2014, United States Department of Agriculture, Office of the Chief Economist, accessed 2015, http://www.usda.gov//oce/foodwaste/usda_commitments.html.

[38]"Waste," Walmart, accessed 2015, http://corporate.walmart.com/global-responsibility/environment-sustainability/waste.

[39]"Walmart Announces New Commitment to a Sustainable Food System at Global Milestone Meeting," October 6, 2014, http://news.walmart.com/news-archive/2014/10/06/walmart-announces-new-commitment-to-a-sustainable-food-system-at-global-milestone-meeting.

[40]"Smaller Portions, Big Benefits," *Manage My Restaurant, National Restaurant Association,* accessed 2015, http://www.restaurant.com/Manage-My-Restaurant/Marketing-Sales/Food/Smaller-portions,-big-benefits.

[41]"Principles of Healthy, Sustainable Menus," The Culinary Institute of America, accessed 2015, http://www.menusofchange.org/images/uploads/pdf/CIA-Harvard_Menus_of_Change_2014_Principles_%28from_report%291.pdf.

[42]Henry Samuel, "Sacré Bleu! French Restaurateurs Asked to Hand Out Doggy Bags," *The Telegraph,* April 15, 2015, http://www.telegraph.co.uk/news/worldnews/europe/france/11537970/Sacre-bleu-French-restaurateurs-asked-to-hand-out-doggy-bags.html.

[43]Mark Bittman, "How to Feed the World," *The New York Times,* October 14, 2013, http://www.nytimes.com/2013/10/15/opinion/how-to-feed-the-world.html?pagewanted=all&_r=0.

CHAPTER 11: URBANIZATION: FOOD FOR THE CITY

EVERY SINGLE DAY 180,000 PEOPLE AROUND THE WORLD MOVE TO A CITY.[1]

This flood of humankind from the suburbs and country-side means that by midcentury, two-thirds of us—some 6 billion people—will live in urban areas.[2] This *could* be very good news. Planned thoughtfully, cities attract talent, create economic opportunity, encourage investment, support community, spread ideas and build wealth. They make access to education and health care affordable to large numbers of people. And they can provide housing, power, water and sanitation for a dense population in ways that are far more sustainable than for a widely dispersed suburban or rural population.

One estimate suggests that in the generation before 2005, urbanization lifted one-half billion people from extreme poverty.[3] Another forecasts that cities will become our engines of wealth, producing 86 percent of global GDP in 2025.[4] If humankind's new "built environment" is people-centered and sustainable, the 21st century's extraordinary urban transformation will drive the prosperity and well-being of the world.

Of course, there are a few hurdles to overcome. A safe and nutritious food supply is just one of the fundamental features required for a healthy city. Given all that we know, from the intense competition for land and water to the impact of climate change, how do we ensure food security for billions of new city dwellers? In many ways this is a conversation that is already late in starting. "Among the basic essentials for life—air, water, shelter and food," one expert in urban planning noted, "planners traditionally addressed them all with the conspicuous exception of food."[5]

The dramatic surge in global food prices in 2007 and 2008 changed everything. Millions of urban residents who considered themselves "food secure" were shocked by sudden high prices that sometimes led to violence. Cities in 20 nations were torn by riots,[6] including one-third in middle- and high-income countries.[7] North and South, rich and poor, residents in cities around the world began to wonder if their supply of food was as secure and stable as they had once believed.

In the last decade, these jolts to urban food systems have shifted the conversation. What began in the 20th century as a dialogue about enhancing food security has become in the 21st century an uneasy conversation often centered around *urban* food security.

WHAT'S SO DIFFERENT ABOUT A CITY?

At a basic level, when people move from rural areas to cities, there are just as many mouths to feed but fewer people producing food. When 180,000 move every day, the balance shifts quickly and dramatically. Consumers in urban areas become distant from their food sources and disconnected from the people and systems that are producing them. For instance, when

CONSUMERS IN URBAN AREAS BECOME DISTANT FROM THEIR FOOD SOURCES

a team from Columbia University studied the New York City food system, one of the most established, flexible and robust in the world, they concluded:

> City dwellers obtain food from a variety of locations: the local bodega or grocery store, the Halal truck on the corner, the office or school cafeteria, or the Indian restaurant down the street. Rarely does one consider where this food comes from, how it gets here, or what might have happened along the way.... The majority of New Yorkers know only that food eventually gets to where it needs to be: on their plates.[8]

This is the kind of disconnect—especially when coupled with abundance and low prices—that can lead to massive food waste.

New urbanites also find themselves in a cash-based market economy very different from life on a farm; when times are hard, many can no longer rely on food growing outside their door for sustenance.

As we know, cities compete aggressively for the same kind of open land and available freshwater that farms do. Urbanization almost always triumphs, reducing prime agricultural land. This pushes farmers farther from their markets while stressing logistics and transport links that bring food to market.

Even if farmers in the shadow of an expanding city are not forced to sell, they sometimes convert from staple crops to more profitable businesses like nurseries or vineyards to serve the growing middle class. This same middle class also demands greater access to vegetables, fruit, meat, eggs and dairy products. In short, the expansion of urban areas means that more and more perishable products must travel greater distances, increasing the risk of food loss.

Time, tempo and space in a city are also different than in a rural setting. Wage earners often work long hours and sometimes have long commutes. Consequently, city residents tend to eat many of their meals outside the home.[9] Sometimes this also reflects the lack of space for meal preparation within the household or lack of food storage and household refrigeration. Street foods, despite their safety risk and lack of nutrition, remain an attractive option for many urban residents. For example, in booming Nigerian cities, residents spend half their food budget on street foods.[10]

The reliance on cheap and convenient food means that both rich and poor cities face overnutrition, a measure of overweight and obese adults. In 2010, global overnutrition exceeded 1 billion people.[11]

This is also a good reminder that while cities create opportunity and wealth, they can also perpetuate poverty. One-third of the world's population lives in slums and informal settlements.[12] Urban poor are highly dependent on affordable food; high prices translate directly into food insecurity, hunger and malnutrition. Average South Africans spend 25 percent of their budget on food–unless they live in the slums of Cape Town, in which case they spend 53 percent of household income to feed their families.[13]

Cities, both rich and poor, have also become notorious

for their "food deserts," defined as urban neighborhoods that do not have access to fresh, healthy and affordable food. The USDA estimates that 23.5 million Americans live in food deserts.[14] Washington, D.C., is an example of one such wealthy yet vulnerable urban area. Eight census tracts in the city qualify as food deserts. Of the 520 food retailers in D.C., 88 percent do not offer any fresh produce.[15] Overall, some 13 percent of D.C. households are food insecure.[16]

As we ponder what's different about a city, it's worth noting that not all are created equal. Thirty-six cities are now classified as "mega-cities" with populations over 10 million.[17] But nearly half of the world's 3.9 billion urban dwellers reside in relatively small settlements with fewer than 500,000 inhabitants. There is no single urban landscape and therefore no single urban food model. Strategies for food production, delivery and security must be customized to local settings.

A TALE OF THREE CITIES

The story of urbanization in Latin America provides a glimpse into food security challenges the entire world will likely face in the coming decades. After 60 years of sometimes haphazard growth, about 80 percent of the region's population now lives in towns or cities. Forecast to be 87 percent by 2050,[18] Latin America and the Caribbean is already the most urbanized region in the world. The United Nations Human Settlements Programme concluded that it had been a largely positive process despite violence, environmental degradation and deep social inequality.[19] Per capita income has nearly tripled since 1970[20] while Latin America's 16,000 cities generate two-thirds of the region's GDP. However, these cities also include 111 million people living in slums and shanty towns.[21]

Three Brazilian cities—Manaus, São Paulo and Belo Horizonte—help describe both the challenges and victories possible as the world seeks to feed its cities in the 21st century.

It's largely impractical to drive from Manaus, a major river port in Brazil's Amazon forest, southward to São Paulo, the largest city in the country. Manaus is bounded by water to the south and has just one major paved road, which stretches north to Venezuela and beyond.[22] Brazil itself has several major highways and about 1.2 million miles of total roads, but only 10 percent are paved. As one observer who visited the city wrote, "there are still no roads to Manaus."[23]

Nonetheless, the farmers and merchants of Manaus produce nutritious foods to sell to São Paulo's growing population. These include perishable fruits like açaí and cupuaçu, sometimes called "superfruits" for their vitamin and mineral content. These are precisely the kinds of foods that address malnutrition and can help eliminate food deserts in a city. Food shipments from Manaus, which may also include squash and watermelon, inevitably begin by water along the Amazon River and Rio Madeira. These travel some 200 miles in three days to Porto Velho. Once landed, the perishables are transferred to trucks, where

Brazilian transport drivers can face conditions ranging from flooded country roads to crushing urban traffic. São Paulo itself is one of the most heavily congested cities in the world; traffic jams average 112 miles and extend to 183 miles with bad weather or on weekends. One expert notes that this heavy traffic takes its toll on the cost of doing business: "If you have a truck and this truck cannot make more than six to eight deliveries instead of 15 or 20, you need two trucks, so everything becomes more expensive."[24]

Despite these obstacles, there are farmers who successfully move perishable product from Manaus to São Paulo. However, as one observer wrote, growers in northern Brazil attempting to supply large cities like São Paulo "produce crops that have to overcome leaps and hurdles for delivery, yet can only be cultivated seasonally and in limited amounts from the forest."[25] The lack of an affordable, dependable cold chain puts both rural farmers and urban residents at risk.

While not as large as São Paulo, the city of Belo Horizonte is home to nearly 2.5 million residents.[26] It forms the core of a metropolitan region that comprises urban and rural areas with a total population of more than 5.7 million, making it Brazil's third most populous urban area after São Paulo and Rio de Janeiro.[27]

In 2009, Belo Horizonte received the "Future Policy Award" from the World Future Council for creating "the world's most comprehensive policy that tackles hunger immediately and secures a healthy food supply for the future."[28] In fact, some call Belo Horizonte "the city that ended hunger."[29] Leaders in Belo Horizonte see access to healthy food as a simple measure of social justice. They applied a series of basic concepts, including tying local producers directly to consumers, educating the population about food security and good nutrition, diversifying agricultural production and job creation, and integrating logistics and supply chain throughout their entire food system.[30] This meant offering local family farmers attractive public space to sell to urban consumers. The city also supported 38 community gardens and three large commercial gardens.[31] Some 34 markets were established throughout the city where some two-thirds of items are sold at low prices and the other third at market prices.[32]

As a result of these and other measures, 25 percent fewer residents of Belo Horizonte live in poverty and 75 percent fewer children under 5 years old are hospitalized for malnutrition. Some 40 percent of the city's residents report frequent consumption of fruits and vegetables against a national average of just 32 percent.[33]

It's important to note that community gardens provide genuine if somewhat limited benefits. Even in Belo Horizonte, the city's food system handles about 45,000 metric tons of food annually while urban agriculture contributes just 50 metric tons.[34] Urban gardens mean stronger social networks, additional income and nutritious foods, but are supplementary to the city's much larger food system footprint.

A United Nations summary of rural transport in Latin America concludes that "transportation is fundamental for the development of human activities, especially production and trade-related activities, including the production and trade of agricultural products."[35] The long, complicated trek from Manaus to São Paulo proves that. As roads improve and refrigerated transport becomes more affordable, the U.N. report notes that cold chain solutions are also becoming popular as "hub" locations that can be shared by smallholder farmers:

> A common solution to transport and logistical barriers in many countries has been to set up collection plants and/or pre-processing plants, usually in association with cooling, refrigeration and/or freezing facilities. These are designed to ensure the safe transport of perishable food products from farm to market, at the same time facilitating physical and commercial integration of the various supply chain components. This has frequently proven an effective measure. ...[36]

This tale of three cities in Brazil highlights many challenges arising in a world where millions of people depend upon nutritious but perishable foods from distant farmland. Roads must be passable. Refrigeration must be available and affordable. Handoffs from ship to truck and truck to warehouse must be coordinated. Information and communications need to be robust along the entire food chain. And once having arrived in a densely populated city, food products must be stored and transported efficiently so that weeks of successful cross-country travel are not lost in the "last mile" of delivery.

In a world where cities are growing and agricultural production is becoming more distant, building and

STRONG LINKAGES BETWEEN RURAL PRODUCTION AND URBAN CENTERS ARE CRITICAL TO URBAN FOOD SECURITY

maintaining strong linkages between rural production and urban centers is critical to urban food security.

OTHER URBAN FOOD STRATEGIES

Some observers believe we have been thinking about the problem of urban food security incorrectly. The divide between urban and rural is "artificial and counter-productive"; the two sectors are part of a larger integrated system.[37] Consequently, cities like Belo Horizonte are not just consumers of food but living engines that help power and sustain rural agriculture.

Amsterdam's food strategy is an excellent example designed to create a stronger ecosystem throughout its adjacent agricultural region. Initiatives involve enhanced local food production, improved eating habits, and a countryside able to meet the food and recreational demands of city residents. "With almost 40 percent of Amsterdam's ecological footprint caused by the provision of food," the city's planning document states, "there was plenty of reasons for the city of Amsterdam to develop a strategy of healthy, sustainable, regional food chains, with special emphasis on urban-rural relationships."[38] Essential to this strategy, the city and its neighboring provinces have mapped supply and demand of regional products in order to strengthen the rural-urban link and improve delivery of fresh, local foods.

Likewise, the city of Baltimore, Maryland, has

developed a sustainability plan that increases the urban land being cultivated and improves the quality and quantity of food available at retail. The city emphasizes identifying the best distribution strategies for existing food networks and filling gaps where networks do not exist.[39]

The work being done by U.K.-based supermarket company Sainsbury's to reduce food waste is also indicative of the ways that urban centers can contribute to sustainable rural agriculture. "Recovery of food losses for increased food security coupled with biomass recovery for energy production and waste conversion to farm inputs should be an integrated strategy, for which cities and towns can be a proving ground," an FAO study concludes. This "greening the economy with agriculture" is a strategy that uses organic urban waste in compost production which can be returned to crop growth.[40]

In China, the green building movement has led to ingenious solutions linking urban processes with food production. Former Vice Minister and now State Councilor Qiu Baoxing sees energy-efficient, green buildings of the future that also source food at the building site, such as planted walls, and even on-site fish farms. This is strong recognition that urbanization and food are inextricably linked in a country where 300 million people—the size of the entire United States—will move to a city by 2030.

The concept of establishing food hubs is an effective strategy long used by cities to provide a central point for the gathering of perishable produce from the countryside.

Hubs take the burden of last-mile delivery from food producers and hand it to players who are experts in urban logistics. In Philadelphia, the Greensgrow Farms' food hub operation has been active in food distribution for almost two decades. In St. Louis, Fields Foods' Neighborhood Grocery is creating a metro market for local produce, meats and dairy. And Hunts Point Produce Market in The Bronx has long been the central food hub of New York City. Some 22 percent of the region's fruits and vegetables flow through this 329-acre center, known as the "Grand Central Station of Broccoli."[41] Hunts Point is the start of an inner-city cold chain that is complex and highly effective, one whose mission is to ensure that the farmers' labors do not go to waste.

Cities can also play a prominent role in ensuring food safety, a topic that has received intense scrutiny in recent years. Sometimes this means a more careful regulation of street food to ensure that it meets health and safety regulations. Sometimes it means increased vigilance over rapidly growing restaurants and food retailers, or establishing safety protocols around water and food waste. Of particular note in rapidly expanding cities is the presence of urban livestock. This is not a few animals intended to supplement a family's diet but rather the existence of intensive livestock production amid congested city dwellings. The benefits of having this protein so readily available are attractive but the cost is potentially steep. "Predictably," one report says, "placing large, dense human populations in close proximity to large, dense livestock populations brings both public health and environmental hazards."[42]

Food safety is one special urban challenge for which infrastructure improvements and cold chain development can be pivotal. By establishing a robust cold chain, city dwellers can enjoy the benefits of fresh meat while keeping people and livestock separate and healthy.

ONE GIANT FOOD ENGINE

The emphasis for global food security in the 21st century is increasingly focused on the food security of cities. More and more people live farther from their food sources. They no longer play a role in its production. People flocking to cities create enormous, demanding markets. Supply chains grow longer and more complex. Meanwhile, many of the fastest-growing cities in the world are in tropical and semitropical locations at special risk to climate change.

Healthy cities embrace locally produced foods, robust linkages between rural and urban settings, and even globally sourced food—which may be more sustainable once all dimensions of travel, water and energy are calculated. They place a premium on nutritious produce, some of which can be grown in urban gardens, but much of which is dependent on effective cold chains. Food security requires city leaders and residents to view themselves not as buildings and blocks, but rather an expansive, complex web that incorporates urban and rural areas. A city with true food security will begin to resemble one giant food engine—where residents are expected not just to consume, but to play a role in sustaining and powering the entire system.

[1]"News and Broadcast," The World Bank, 2013, http://web.worldbank.org/WBSITE/EXTERNAL/NEWS/0,,contentMDK:20149913~menuPK:34457~pagePK:64003015~piPK:64003012~theSitePK:4607,00.html.

[2]"World's Population Increasingly Urban With More Than Half Living in Urban Areas, United Nations Department of Economic and Social Affairs, July 10, 2014, http://www.un.org/en/development/desa/news/population/world-urbanization-prospects-2014.html.

[3]*Unlocking American Efficiency: Economic and Commercial Power of Investing in Energy Efficient Buildings,* United Technologies and Rhodium Group, May 2013.

[4]*The Future of Urban Mobility: Towards Networked, Multimodal Cities of 2050,* Arthur D. Little Future Lab, No. 1: Future of Urban Mobility, http://www.adlittle.com/downloads/tx_adlreports/ADL_Future_of_urban_mobility.pdf

[5, 29]Kevin Morgan, "Feeding the City: The Challenge of Urban Food Planning," International Planning Studies, Vol. 14, No. 4, November 2009, http://www.tandfonline.com/doi/pdf/10.1080/13563471003642852, 341, 346.

[6, 7, 37, 40]"Food, Agriculture and Cities: Challenges of Food and Nutrition Security, Agriculture and Ecosystem Management in an Urbanizing World," FAO Food for Cities multi-disciplinary initiative position paper, Food and Agriculture Organization of the United Nations, 2011, http://www.fao.org/3/a-au725e.pdf, 3, 13, 16, 21.

[8]"Understanding New York City's Food Supply," Prepared for New York City Mayor's Office of Long-Term Planning and Sustainability by Columbia University May 2010, http://mpaenvironment.ei.columbia.edu/files/2014/06/UnderstandingNYCsFoodSupply_May2010.pdf, 6.

[9, 11, 13]"Feeding Cities: Food Security in a Rapidly Urbanizing World: Conference Report," Penn Institute for Urban Research, March 2013, http://penniur.upenn.edu/uploads/media/Feeding-Cities-Report.pdf, 17, 11, 10.

[10]Cecilia Tacoli et al, "Urban Poverty, Food Security and Climate Change," International Institute for Environment and Development, Human Settlements Working Paper No. 37, March 2013, http://pubs.iied.org/10623IIED.html, 18.

[12]"Food for the Cities," Food and Agriculture Organization of the United Nations, accessed 2015, ftp://ftp.fao.org/docrep/fao/012/ak824e/ak824e01.pdf.

[14]"Food Deserts," United States Department of Agriculture, Agricultural Marketing service, accessed 2015, http://apps.ams.usda.gov/fooddeserts/fooddeserts.aspx.

[15, 16]Sabine O'Hara, "Food Security: The Urban Food Hubs Solution," Resilience.org, April 22, 2015, http://www.resilience.org/stories/2015-04-22/food-security-the-urban-food-hubs-solution.

[17]"Megacity," Wikipedia, accessed 2015, http://en.wikipedia.org/wiki/Megacity.

[18]"UN-Habitat Global Country Activities Report: 2015 – Increasing Synergy for Greater National Ownership," UN Habitat for a Better Urban Future, http://unhabitat.org/books/un-habitat-global-country-activities-report-2015-increasing-synergy-for-greater-national-ownership/, 53.

[19, 20, 21]"State of Latin American and Caribbean Cities 2012," UN Habitat for a Better Urban Future," August 2012, http://unhabitat.org/books/state-of-latin-american-and-caribbean-cities-2/, vii, 37, 61.

[22]Mario Osava, "BRAZIL: Bridge to Drive Urban Growth in Heart of Amazon," May 26, 2010, IPS-Inter Press Service, http://www.ipsnews.net/2010/05/brazil-bridge-to-drive-urban-growth-in-heart-of-amazon/.

[23]"England v Italy: Are They Painting the Manaus Pitch Green?", The Independent, accessed 2015, http://www.independent.co.uk/sport/football/international/england-v-italy-manaus-in-manic-race-to-be-ready-for-opener-9533539.html.

[24]Paulo Cabral, "São Paulo: A City With 180km Traffic Jams," BBC News, September 25, 2012, http://www.bbc.com/news/magazine-19660765.

[25]Emma Young, "From North to South: How the Amazon Feeds São Paulo," *The Miracle of Feeding Cities,* September 10, 2014, http://miracleoffeedingcities.com/from-north-to-south-how-the-amazon-feeds-sao-paulo/.

[26]"Belo Horizonte," *Wikipedia,* accessed 2015, http://en.wikipedia.org/wiki/Belo_Horizonte.

[27, 31, 34]"Belo Horizonte," Growing Greener Cities: Urban and Peri-urban Agriculture in Latin America and the Caribbean, Food and Agriculture Organization of the United Nations, accessed 2015, http://www.fao.org/ag/agp/greenercities/en/GGCLAC/belo_horizonte.html.

[28, 30, 33]"Celebrating the Belo Horizonte Food Security Programme," World Future Council, 2009, http://www.worldfuturecouncil.org/fileadmin/user_upload/PDF/Future_Policy_Award_brochure.pdf.

[32]Frances Moore Lappe, "The City That Ended Hunger," YES! Magazine, February 13, 2009, http://www.yesmagazine.org/issues/food-for-everyone/the-city-that-ended-hunger.

[35, 36]"Rural Transport of Food Products in Latin America and the Caribbean," FAO Agricultural Services Bulletin 155, Food and Agriculture Organization of the United Nations, 2008, ix, 3.

[38]"Amsterdam Food Strategy," The Netherlands, accessed 2015, http://ec.europa.eu/regional_policy/archive/conferences/urban_rural/2008/doc/pdf/6a_iclei_amsterdam.pdf.

[39]"The Baltimore Sustainability Plan," Baltimore City Council, April 2009, http://www.baltimoresustainability.org/sites/baltimoresustainability.org/files/Baltimore%20Sustainability%20Plan%20FINAL.pdf.

[41]Tove Danovich, "The One Hundred Acre Produce Market in the Bronx: Have You Heard of Hunts Point?", The Miracle of Feeding Cities, August 27, 2014, http://miracleoffeedingcities.com/a-one-hundred-acre-produce-market-in-the-bronx-have-you-heard-of-hunts-point/.

[42]2014-2015 Global Food Policy Report, Washington, D.C.: International Food Policy Research Institute, 2015, http://www.ifpri.org/sites/default/files/publications/gfpr20142015.pdf, 44.

CHAPTER 12: OTHER STRATEGIES TO FEED THE WORLD

SOMETIMES IT'S GOOD TO BE WRONG.

When 19th-century scholar Thomas Malthus concluded that the world's population would one day outstrip its food supply, he failed to anticipate the amazing 20th-century scientific advances in agriculture.

Now as we consider ways to reduce food loss and waste, what revolutionary ideas might we anticipate will help feed the world and heal the planet? Perhaps some solutions will provide higher crop yields without the need for more land, water or other resources. Others might involve ways to insulate farmers from drastic weather events or teach consumers new ways to buy and prepare food. Some might involve underutilized or entirely new sources of food.

For example, we could all eat bugs.

LARGE-SCALE INSECT FARMING

It's actually called "entomophagy"—the consumption of insects by humans—and it's a practice that dates back presumably to the beginnings of humankind, or at least the moment when early man got really hungry. In fact, bugs are just one kind of insect; today, about 1,900 insect species supplement the diets of some 2 billion people around the world.[1]

What's now emerging as a viable industry is large-scale insect farming. The pet food and fish bait industries have been dabbling in this for years. The development of such an industry would have a number of advantages. Insects are everywhere. They can be gathered and cultivated by almost anyone, making entrepreneurial ventures possible even in the poorest of countries. Insects reproduce quickly and convert feed efficiently, at about the rate of poultry. They require less water and land than conventional livestock and have lower carbon emissions.

In fact, insects are comparable to soy in protein—yet require significantly less land. Trials have determined that the same hectare of land that can produce just under a ton of soy protein annually could produce 150 tons of insect protein.[2]

There are hurdles, of course, including the fact that there are few tested models for mass production. Insects need to be safe and disease-free. New regulations would have to be passed in many countries. And there's the need to make this new food source attractive to consumers, suggesting education and trial in many markets. However, even in places where humans choose not to consume them directly, insects might still be used as feedstock for aquaculture and poultry, supplementing traditional soy, grains, maize and fishmeal.[3] One company in France is seeking to launch the first fully automated, large-scale insect production facility, capable of producing 10,000 tons of dried protein annually.[4] This kind of innovative initiative will take pressure off conventional agriculture, while creating new business opportunities for entrepreneurs around the world.

If protein bars[5] or tortilla "chirps"[6] made from cricket flour don't sound appetizing, there's another, even bigger opportunity to develop large-scale, sustainable and nutritious animal protein. And this one is already helping to feed the world.

THE RISE OF AQUACULTURE

The ocean has been good to humankind for millennia, yet remains one of our greatest underutilized resources. Covering 70 percent of the Earth, it provides just 17 percent of our animal protein[7] and only 2

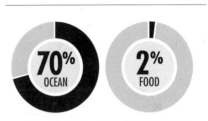

COVERING 70 PERCENT OF THE EARTH, THE OCEAN PROVIDES JUST 2 PERCENT OF OUR TOTAL CONSUMED FOOD

percent of our total consumed food. The fishing industry plays a vital role in the world economy, employing millions of people and comprising 10 percent of total agricultural exports.[8]

Unfortunately, humankind has not always been good to the ocean. Overfishing has taken an enormous toll on sea life. Almost 29 percent of global wild fish stocks are being exploited at unsustainable levels.[9] Each year more fishing ships burning more fossil fuel are deployed even though total global capture has been stagnant since the 1990s.[10] Meanwhile, carbon emissions—including the 3.3 billion metric tons from our mountain of food waste—are altering the sea's chemistry. In just the last 200 years, Earth's oceans have experienced a 26 percent increase in the level of acid, a rate of change unlike anything seen in the fossil record over the last 65 million years.[11] The threat to marine food webs, the building blocks of life in the ocean, is alarming. So too is the risk to global food security. A United Nations report concludes that "if CO_2 emissions continue at the same rate, ocean acidification will have a considerable influence on marine-based diets for billions of people worldwide."[12]

Ocean acidification is perhaps the best example of the planet's interconnected carbon cycle. The more

CO_2 ↑ OVER 200 YEARS = 26% INCREASE IN OCEAN ACIDIFICATION

carbon we emit from buildings, transportation, wasted food and other sources, the more CO_2 is absorbed by the oceans—and the more current and future sources of food are put at risk. Philippe Cousteau Jr., explorer, journalist, philanthropist and grandson of oceanic pioneer Jacques Cousteau, captures the challenge so well when he says, "The oceans are a proxy for the problems we're trying to solve."[13] Aquaculture—the cultivation of aquatic animals and plants—may provide one such solution. Already a booming global business, it is poised to become the next (and perhaps last) large-scale, sustainable industry capable of producing nutritious animal protein. Despite stagnation in the wild fish catch, aquaculture has increased global fish production 3.2 percent annually, double the growth rate of the population.[14] Today there is more farmed fish produced globally than beef.[15] And fish just happens to be among the most efficient proteins on the planet, requiring a pound of feed to produce a pound of farmed fish. That compares with 2 pounds of feed for a pound of chicken and about 7 pounds of feed for beef.

Of course, aquaculture is not a panacea. It must be designed safely and sustainably from the start. It must find ways to control disease while reducing the use of antibiotics. It must be sensitive to land use. It cannot create new pollution. While more efficient than traditional animal husbandry, it still requires vast amounts of feed. And seafood is highly perishable and dependent upon a robust cold chain for all but the shortest distribution links. However, the industry is learning fast and taking advantage of lessons from conventional livestock breeding. For example, in Norway, zoning laws for salmon producers have reduced concentration and the threat of disease.

Aquaculture has an enormous upside for feeding the world, reducing poverty and enhancing food security. It stands to benefit regions of the world where hunger and malnutrition are most acute. Already Thailand and Vietnam are global leaders in supplying species like shrimp and tilapia. Asia is expected to comprise nearly 90 percent of all new global production, with China alone responsible for half.[16] There is also great promise in Africa for the growth of aquaculture.

However, whether fish are caught wild or produced from aquaculture, the global consumer can play a critical role in the health and well-being of the seafood industry. Just 10 species account for one-quarter of all wild capture, and most of these are fully fished or overfished. In the U.S., just four species—shrimp, tuna, salmon and tilapia—comprise nearly 70 percent of all seafood consumed.[17] Most other species are dismissed as "trash fish" despite many being nutritious, tasty and plentiful. That means the battle for sustainable seafood must be waged plate by plate.

Barton Seaver is a National Geographic Fellow, an award-winning chef, director of the Sustainable Seafood and Health Initiative at Harvard's School of Public Health, and a leader in the movement to design sustainable food systems. His specialty is the healthy and sustainable sourcing of seafood. Chefs "have a huge influence on the guiding hand of natural selection," Seaver says. "We are the ones who popularized bluefin tuna, sea bass and salmon. We also have a chance to change this by using more environmentally sustainable species on our menus instead. We can harm or we can heal. We also have the opportunity to be educators and advocates."[18]

If consumers are willing to experiment, they could soon sample menu items with names like snakehead, black drum, lingcod, scup, porgy and grunt. Seaver often says, "For too long we've been telling the ocean what we want from it, instead of the ocean telling us what it can give." Perhaps enterprising chefs will collaborate with enterprising marketers: In 1977, the Patagonian toothfish was renamed "Chilean Sea Bass" by a resourceful fish wholesaler[19] and is now proudly featured in many restaurants and supermarkets.

Investments in aquaculture and consumer education

will not only generate vast new sources of healthy protein, but will also help heal the oceans. The FAO estimates that rebuilding overfished stocks could generate an additional $32 billion in global annual revenue.[20]

AN OLD IDEA MADE NEW

In a very real sense, agroecology—defined as the application of ecology to agriculture and food systems—is the way humankind used to farm. When designed well, agroecosystems are sustainable, rely on minimum artificial inputs, and manage pests and diseases through internal regulating mechanisms.[21] Over a century old as a discipline, agroecology gained new stature in 2010 when the United Nations Special Rapporteur on the Right to Food recommended it as a viable approach to achieving food security.[22]

For example, on farmland in Niigata, Japan, wheat and barley were sewn amid watermelons. These grains did not compete with the watermelon, conserved soil moisture and reduced weeds.[23] After harvest the wheat and barley became mulch, enriching the soil for the next planting. Agroecology took on a different form in the Andhra Pradesh state of India. There, smallholder farmers combined compost, dung and nitrogen-fixing plants to improve soil quality. The result was reduced agricultural inputs and increased yield by several hundred pounds per acre for a variety of crops.[24]

Sometimes plants that drive away pests are inter-planted with the primary crops. Sometimes agroecology is as simple as building stone barriers to reduce rainwater runoff.

The twin emphases of agroecology are sustainability and yield. Agroecology includes a perspective that looks beyond the field to the larger ecosystem and community. Often it can provide an alternative to intensive, mono-culture farming in regions of the world that lack the inputs or expertise for such practices. And—despite being knowledge-intensive, customized by location and need, and sometimes perceived as risky by traditional small-holder farmers—agroecology is being driven by practical success. A 2011 study reported that 40 initiatives in 20 countries involving 10.4 million farmers demonstrated substantially improved crop yields with reduced carbon emissions, pesticide use and soil erosion.[25]

Proponents of agroecology see success in farming and reduced poverty as two tightly bound phenomena. This turns the old question on its head, asking not how to feed the world but how the world can feed itself.[26]

PROMISE AND CONTROVERSY

In a poor region of Africa where both malnutrition and pesticide poisoning are common, a smallholder farmer grows eggplant on his 1 acre of land. He doesn't use pesti-cides, yet yields are up and so is his income. This is the result of a variety of pest-resistant eggplant made possible by transgenetic modification. In other words, a gene transferred from a soil bacterium into the eggplant variety creates a protein that is nontoxic to humans but deadly to the pests that try to destroy his crops.

No one can argue that this new variety of eggplant is feeding this farmer's family and lifting him from poverty.[27] At the same time, such "genetically modified (GM) foods" are banned in many countries around the world, have led to protests and are part of a serious, ongoing public debate.

GM crops were first released in the 1990s. Since then, they have been adopted by 17 million farmers in 20 developing and eight developed countries. GM crops have been planted on 11 percent of the world's arable land. Half of the cropland in the U.S. is planted with GM crops including 90 percent of all corn, cotton and soybeans grown. Americans have been eating GM products for almost 20 years.[28]

Globally, GM crops grew one hundredfold from 1.7 million hectares in 1996 to 170.3 million hectares in 2012—the most quickly adopted crop technology in recent history.[29]

GM foods on the international market have passed rigorous safety tests. Consumption by the general public to date has shown no ill effects on health. And yet, a recent survey of the American Association for the Advancement of Science showed 88 percent of members agreed it was safe to eat genetically modified foods while only 37 percent of the public did. This is a greater gap than the difference between science and the public regarding climate change or childhood vaccinations.[30]

The public has genuine concerns. One is the fear that gene transfer in GM foods could pass to the cells of the human body. Another is the possibility of GM plants "outcrossing" or migrating into the genes of conventional crops. There are also questions about the validity of safety assessments when the impact of GM foods on humans might be long-term. And, not unlike superbugs created by antibiotics, GM crops can also have a profound impact on their environment: Just as crops adapt to pests, pests adapt to crops. This suggests that as scientists strive to create supercrops, Nature strives just as surely to create superpests and superweeds.

At a minimum, consumers in many countries remain adamant that GM food be clearly labeled so that they can make informed choices.

The positive impact of GM foods on global food security could be enormous, but science and industry must overcome consumer questions before such foods are widely embraced and globally consumed.

As scientists seek to feed the world, they are undertaking other kinds of revolutionary genetic work. The deciphering of the genomic content of crop species allows scientists to manipulate genes within the same species to create superior crops.[31] This is a huge improvement over the slow hybridization of the past, and advances can be stunning. For instance, scientists announced recently that they had engineered rice plants to boost their ability to capture CO_2. The result was faster photosynthesis, which might increase yields by 50 percent or require less water and fertilizer to produce the same amount of food.

Likewise, the African Orphan Crops Consortium has begun work with genome sequencing that will improve the nutritional content, productivity and climate adaptability of some of Africa's most important food crops. "Next-generation genomics is one of the 'best bets' for sustainably eradicating hunger, malnutrition and poverty," one expert concludes.[32]

BIG DATA MEETS AGRICULTURE

When it comes to revolutionary ideas, few have garnered as much recent press as "Big Data." Defined by consulting firm McKinsey & Company as "datasets whose size is beyond the ability of typical database software tools to capture, store, manage, and analyze,"[33] Big Data seems perfectly suited to a business where season and weather alone are infinitely complicating factors. Accordingly, farm analytics companies are springing up everywhere. Solutions offering "precision agriculture" measure field variability using GPS and satellites. Some services assist farmers in planning seed amount, spacing and depth. Others determine soil productivity and forecast yield. Some package insurance with data analysis to protect against increasingly volatile weather. The ultimate goal is to optimize inputs like seed and fertilizer, compare results over time, and continue to lower costs and improve yields. Ultimately, the farmer wants to plant more seed in high-potential areas and less in low-potential areas—and have the confidence of knowing which is which.

Some Big Data solutions offer a more global perspective. An interactive website launched in 2015 provides a snapshot of food and nutrition security in 22 Arab countries. This is an especially welcome addition after the food shocks of 2007 and 2008. The service is designed to track more than 150 socio-economic and biophysical indicators such as food consumption, malnutrition and disease, helping policymakers enhance food security in the region.[34] Another service provides crop production estimates for 42 crops around the world. This shows where in the world individual crops are cultivated, their production patterns and whether they are irrigated or rain-fed. This helps policymakers optimize food production and meet consumer demand.[35]

A third service is a combination phone and tablet computer designed to help smallholder farmers in developing countries purchase affordable agriculture materials, get better prices for their produce and link more effectively to markets.[36] And finally, sometimes all a revolution takes is leveraging older technology and people in new ways: Countries in Africa and South Asia are now providing important weather data to millions of farmers by mobilizing radio stations, meteorological services, religious groups, agricultural agents and schools, and farmers themselves. For example, in Senegal's Kaffrine District, farmer demand for climate predictions has spurred the growth of an integrated climate forecasting program that today reaches 5,000 farmers and is slated to one day reach 2 million.[37]

NEW IDEAS FOR CONSUMERS

As the world's population grows and urbanization draws more people away from farms and the production of their own food, technology might play a role in spanning the divide. Radio Frequency Identification (RFID) uses a wireless microchip and antenna. Attached to a truck or container, a box, or even a head of lettuce, it can track food products as they move throughout the supply chain. Some RFID tags have temperature and humidity sensors able to report on the particular conditions perishable food encounters on its journey. Such location and condition data can be used to optimize routing, expedite shipments at risk and speed product around the world.

As tags become less expensive and more powerful, they are also able to store data concerning product attributes. Where was a head of lettuce grown? When? Is it organic? How much shelf life should it have remaining? Such data might also reside in a central database and be accessed via a tag and smartphone. This information could be shared along the entire food supply chain, including the consumer. Often thousands of miles apart, farmers and consumers will be able to communicate about food with a transparency that will enhance trust and food security.

Food retailers are also beginning to offer consumers ingenious data-based solutions to help reduce waste. Computerized shelf labels in use today could allow real-time pricing for produce. Those products nearing the end of useful life could be priced lower to spur consumption instead of inevitable waste. Sainsbury's Food Rescue is another simple web tool that helps find recipes to use up food that may otherwise go to waste. A consumer enters the ingredients on hand in his or her kitchen and the smartphone app identifies recipes that use those ingredients.[38] Then there are ideas designed to simply improve clarity around food decisions. One is to rationalize or even scrap "best before" labels on long-life products. In the Netherlands, for example, about 15 percent of food is wasted because of misunderstandings about product labeling.[39] Those faced with an array of "use before," "best if used by," "sell by" and "open date" codes know that rationalization of terms would reduce confusion and likely reduce waste.

Finally, one of the most revolutionary ideas may be one of the simplest. Retailers are now testing the sale of products, including perishable foods, from bulk bins with little or no packaging. This allows shoppers to reuse their own containers and to purchase precisely the amount they require, cutting down on waste.[40] For those consumers who forget their containers, these "zero-waste" grocers will also sell long-term green packaging.[41] While these efforts are still in the early stage, observers are watching to see if food waste can be reduced.

[1, 3]"The Contribution of Insects to Food Security, Livelihoods and the Environment," Food and Agriculture Organization of the United Nations, 2013, http://www.fao.org/docrep/018/i3264e/i3264e00.pdf.

[2]Lauren Zanolli, "Insect Farming Is Taking Shape as Demand for Animal Feed Rises," MIT Technology Review, August 20, 2014, http://www.technologyreview.com/news/529756/insect-farming-is-taking-shape-as-demand-for-animal-feed-rises/.

[4]Ynsect, http://www.ynsect.com/.

[5]Exo, https://www.exoprotein.com/.

[6]Six Foods, http://www.sixfoods.com/#products.

[7, 8, 9, 14, 20]"The State of World Fisheries and Aquaculture: Opportunities and Challenges," Food and Agriculture Organization of the United Nations," Rome, 2014, 4, 7, 3.

[10]Siwa Msangi and Miroslav Batka, "The Role of Fish in Global Food Security," Washington, D.C. International Food Policy Research Institute, 2015, http://ebrary.ifpri.org/cdm/ref/collection/p15738coll2/id/129078, 62.

[11, 12]Carol Turley et al, "Environmental Consequences of Ocean Acidification: A Threat to Food Security," UNEP, 2010, http://www.unep.org/dewa/Portals/67/pdf/Ocean_Acidification.pdf, 2, 1.

[13]World Cold Chain Summit to Reduce Food Waste, London, England, November 20, 2014.

[15]Joel K. Bourne, Jr., "How to Farm a Better Fish," National Geographic, June 2014, http://www.nationalgeographic.com/foodfeatures/aquaculture/, 93.

[16]Siwa Msangi and Miroslav Batka, "The Role of Fish in Global Food Security," Washington, D.C. International Food Policy Research Institute, 2015, http://ebrary.ifpri.org/cdm/ref/collection/p15738coll2/id/129078, 64.

[17]Elizabeth Gunnison Dunn, "Why These Overlooked Fish May Be the Tastiest (and Most Sustainable)," Wall Street Journal, March 20, 2015, http://www.wsj.com/articles/when-it-comes-to-fish-one-chefs-trash-is-anothers-daily-special-1426870428.

[18]Jadda Miller, "Creating a Sustainable World: An Interview With Barton Seaver," Nourishing the Planet, August 10, 2011, http://blogs.worldwatch.org/nourishingtheplanet/creating-a-sustainable-world-an-interview-with-barton-seaver/.

[19]"Patagonian Toothfish," Wikipedia, accessed 2015, http://en.wikipedia.org/wiki/Patagonian_toothfish.

[21]Stephen R. Gliessman, "An Ecological Definition of Sustainable Agriculture," accessed 2015, http://www.agroecology.org/Principles_Def.html.

[22, 25, 26]M. Wibbelman et al, "Mainstreaming Agroecology: Implications for Global Food and Farming Systems," Centre for Agroecology and Food Security Discussion Paper," Coventry: Centre for Agroecology and Food Security," http://futureoffood.org/pdfs/Coventry_University_2013_Maintstreaming_Agroecology.pdf, iii, 2.

[23]"Watermelon Cover Cropping with Wheat and Barley in Niigata, Japan," Case Studies, accessed 2015, http://www.agroecology.org/Case Studies/watermelon.html.

[24]Sylvia Rowley, "In India, Profitable Farming With Fewer Chemicals," The New York Times, April 24, 2015, http://opinionator.blogs.nytimes.com/2015/04/24/in-india-profitable-farming-with-fewer-chemicals/?smprod=nytcore-ipad&smid=nytcore-ipad-share&_r=0.2.

[27, 30]Mark Lynas, "How I Got Converted to G.M.O. Food," The New York Times, April 24, 2015, http://www.nytimes.com/2015/04/25/opinion/sunday/how-i-got-converted-to-gmo-food.html.

[28]Tim Folger, "The Next Green Revolution," National Geographic, National Geographic Society, 2014, 102.

[29]"Global Status of Commercialized Biotech/GM Crops: 2012," ISAAA Brief 44-2012: Executive Summary, International Services for the Acquisition of Agro-Biotech Applications, 2012, http://www.isaaa.org/resources/publications/briefs/44/executivesummary/.

[31]Kevin Bullis, "Supercharged Photosynthesis," MIT Technology Review, Vol. 118, No. 2, March/April 2015, 59.

[32]"Next-Generation Genomics Key to Global Food and Nutritional Security," International Policy Food Research Institute, February 20, 2015, http://www.icrisat.org/newsroom/news-releases/icrisat-pr-2015-media3.htm.

[33]Gil Press, 12 Big Data Definitions: What's Yours?", Forbes, September 3, 2014, http://www.forbes.com/sites/gilpress/2014/09/03/12-big-data-definitions-whats-yours/.

[34]"Website Focusing on Food Security in Middle East Now in Arabic," International Policy Food Research Institute, April 27, 2015, http://www.ifpri.org/pressrelease/website-focusing-food-security-middle-east-now-arabic.

[35]"New Global Crop Data Aid in Food Policy Decisions," International Policy Food Research Institute, April 8, 2015, http://www.ifpri.org/pressrelease/new-global-crop-data-aid-food-policy-decisions.

[36]"GreenPHABLET powered by the GreenSIM," International Policy Food Research Institute, December 29, 2014, http://www.icrisat.org/newsroom/news-releases/icrisat-pr-2014-media40.htm.

[37]"New Study Documents Surge in 'Climate Services" in Africa and South Asia That Allow Farms to See Their Fields in the Future Tense and Adapt to Changes," CGIAR, November 4, 2014, http://ccafs.cgiar.org/news/media-centre/press-releases/climate-change-threatens-food-production-countries-fight-back#.VUUozflVhBd.

[38]"Sainsbury's Food Rescue," B-Reel Creative, accessed 2015, http://www.b-reel.com/projects/digital/case/617/food-rescue/.

[39]Aarthi Rayapura, "Europe Unleashing Full-Scale Attack on Food Waste," Sustainablebrands.com, June 10, 2014, http://www.sustainablebrands.com/news_and_views/business_models/aarthi_rayapura/europe_unleashing_full-scale_attack_food_waste.

[40]Chris Tognotti, "Berlin's 'Original Unpacked' Supermarket Is Packaging-Free, And Greener Than Green," May 30, 2104, http://www.bustle.com/articles/26368-berlins-original-unpacked-supermarket-is-packaging-free-and-greener-than-green.

[41]Krisztina Kupi, "Zero-Waste Supermarket to Open in Belgium," GreenFudge.org, May 15, 2014, http://www.greenfudge.org/2014/05/15/zero-waste-supermarket-open-belgium/.

CHAPTER 13: FOOD WASTE AND PUBLIC POLICY

THE INFLUENCE OF FOOD WASTE REDUCTION IN PUBLIC POLICY HAS ONLY JUST BEGUN.

"The whole waste issue has been very subterranean," says Lauren Chambliss of the Atkinson Center for a Sustainable Future at Cornell University. "We're only now beginning to make the links. Up until very recently very few people outside of academia were looking at food waste in the larger contexts of nutrition, public health, land use and climate change."[1]

Policies that enhance nutrition inevitably rely on perishable foods made widely available by a global cold chain. Consequently, better nutrition demands improved transport and storage. The result is reduced food waste.

In parallel, countries around the world have elevated the issue of food safety in the wake of numerous outbreaks of tainted product and foodborne disease. Both industry- and government-led initiatives have been introduced to improve the welfare of our global food supply. These unavoidably demand better harvest, handling, transport, tracking and storage practices—all of which help to reduce food waste.

The tide is turning. Food waste in the global supply chain has long been the "elephant in the room." With investment, technology and education, a focus on "wasting less" can finally play a role in public policy that is commensurate with its powerful impact on hunger, malnutrition and the environment.

AN INDIRECT APPROACH TO FOOD WASTE

There are numerous public and private initiatives around the globe today intended to improve the global food system. Many address the issue of food waste, but only in a very indirect fashion. For example, in 2014, the government of Pakistan sought to partially guarantee credit for smallholder farmers. This was seen as a way to bring some of that nation's poorest citizens into the market economy and enhance their food security.[2] Farmers might use this new credit to improve their agriculture inputs, enhance their harvest practices or improve transport of their goods. In all cases, food waste would likely be reduced. Similarly, a number of Middle Eastern countries are building their strategic grain reserves to guard against future price shocks. The additional investment and attention paid by these countries to food storage and preservation techniques can help reduce food waste. And Germany announced plans to spend 1 billion euros on food security and rural development through its *One World, No Hunger* program. Initiatives under this program include addressing hunger and malnutrition, preventing famines, developing small-scale farming and establishing centers of innovation.[3] All of these touch upon the issue of food waste.

Nutrition already plays a vital if roundabout role in encouraging reduced food waste. The International Food Policy Research Institute reported that "nutrition shot up to the top of the global development agenda in 2014."[4] One special highlight was the Second International Conference on Nutrition in Rome, during which leaders endorsed 60 action items designed to address the global nutrition problem. Among these were recommendations to produce more fruits, vegetables and appropriate animal-sourced products. Conference leaders also recommended improved "storage, preservation, transport and distribution technologies and infrastructure to reduce seasonal food insecurity and nutrient loss and waste."[5]

The goal was improved nutrition; the outcome should result in an effective cold chain and reduced food waste.

THE ROLE OF FOOD SAFETY

The development of food safety standards can also drive reduced food waste.

Foodborne diseases occur when people eat foods contaminated by pathogens or chemicals. As the world's population grows and food supply chains are stretched, food safety is

THE DEVELOPMENT OF FOOD SAFETY STANDARDS CAN ALSO DRIVE REDUCED FOOD WASTE

increasingly at risk. In fact, as crises have emerged over the last two decades, the safety of our food supply has become an especially visible issue. For example, a scandal involving adulterated cooking oil prompted Taiwan to create a food safety agency in 2014. An outbreak of a pathogen found in meat and raw milk in Australia brought about a restructuring of that country's food safety system. China faced scandal when cadmium was found in rice. An American-owned meat producer in China sold out-of-date and tainted meat to popular restaurants. In Denmark, tainted pork sausages killed 12 people.[6] In the U.S., foodborne illnesses in 2014 included a chicken salmonella outbreak that made almost 650 people ill, tainted cilantro that sickened more than 300 people and tainted bean sprouts that sickened several hundred more, two outbreaks from tainted cheese, and even a chicken dish, served at a business conference, that made 216 people ill.[7] While the numbers are small, the bad publicity that resulted from these outbreaks led to lost revenue and shaken consumer confidence.

In other regions of the world, the numbers are frightening. Foodborne and waterborne diarrheal diseases kill an estimated 2.2 million people annually, most of them children. Diarrhea is the most common symptom of foodborne illness, but other serious consequences include kidney and liver failure, brain and neural disorders, reactive arthritis, cancer and death.[8] In low-income countries, diarrhea from contaminated food and water is a significant issue. However, some of these countries are currently insulated from many foodborne diseases because meats, fish and leafy vegetables are either unavailable or have very short supply chains.[9] This situation suggests that as incomes grow, cities expand, and food supply chains are extended in these developing countries, the risk of foodborne illnesses will grow unless appropriate food safety measures are taken.

The International Food Policy Research Institute believes that it is the rapidly emerging economies, like China and India, which face the most important set of food safety concerns.[10] "They are characterized," the institute's

2014 report concludes, "by rapidly growing demand for the riskiest foods (animal source foods and vegetables), rapidly intensifying agriculture to meet these demands, but lagging food governance systems."[11] Not surprisingly, these also tend to be the economies that offer the greatest opportunity for reducing food waste.

Urbanization complicates food safety. As noted earlier, some countries like China promote livestock production within growing cities, placing animals and people in close, concentrated proximity. Lack of refrigeration in these urban centers adds another risk factor. "Most studies of the farms and wet markets [which sell live animals and unrefrigerated meat and produce] of emerging countries," the report continues, "reveal high levels of pathogens and contaminants."[12] This is a major concern, considering that in places such as Vietnam that lack a modern cold chain, 97 percent of pork is sold in wet markets.

"Many bacterial microbes need to multiply to a larger number before enough are present in food to cause disease," the U.S. Centers for Disease Control (CDC) reports. "Given warm moist conditions and an ample supply of nutrients, one bacterium that reproduces by dividing itself every half-hour can produce 17 million progeny in 12 hours." The CDC concludes, "In general, refrigeration or freezing prevents virtually all bacteria from growing."[13]

However, even in nations with adequate refrigeration and robust food systems, food safety is still an issue. The CDC estimates that one in six Americans (or 48 million people) gets sick, 128,000 are hospitalized and 3,000 die of foodborne diseases[14] by consuming contaminated foods or beverages each year. Foodborne illnesses cost the U.S. economy as much as $16 billion annually.[15]

Strong food safety regulations can make a meaningful difference in the pace of cold chain growth and in the confidence consumers have in the quality and safety of the food they eat. In fact, strong regulations that reinforce best practices are imperative as food production and trade rapidly globalize. A committee of the U.N. has suggested that "closer linkages among food safety authorities is required at the international level"[16] as a way of promoting the integrity of the global food supply chain.

For over a decade, maintenance of the cold chain has been one of the guiding principles of European Union legislation regarding food safety. In its Food Safety Policy, regulators concluded that food producers and distributors should be in temperature compliance with their foodstuffs and maintain the cold chain as a way to enhance food safety.[17] A special section on "Transport" recommends that conveyances used for food transport be kept clean and in good repair, and "where necessary ... are to be capable of maintaining foodstuffs at appropriate temperatures and allow those temperatures to be monitored."[18] The European Food Safety Authority ("committed to ensuring that Europe's food is safe") is an independent organization that provides science-based advice on cold chain and other matters. The EU Health & Consumers Directorate-General sponsors a Working Group on Food Losses/Food Waste that also works to support good regulation and practices.

While Europe has faced several daunting crises, including the rise of "mad cow disease" in the 1980s and PCB contamination in eggs and chickens in 1999, it has responded in a strong, coordinated fashion and today has one of the safest cold chains in the world. And this harmonized approach has had a positive impact on the issue of food waste.

Regulatory pressure on the cold chain is being driven in the U.S. by the Food Safety Modernization Act (FSMA), signed into law by President Obama in January 2011. The law granted the Food and Drug Administration (FDA) new power, including mandatory recall authority, while requiring it to issue new guidance and regulations to industry. There is general agreement by private industry with the broad provisions of the new rule-making, and a willingness to invest ahead of regulations in best practices that improve quality and safety, and protect brand integrity. A senior adviser in the FDA's Center for Food Safety and Applied Nutrition remarked that the FSMA "reflects our understanding that we have one highly integrated food safety system. So we don't have a separate system for domestic and foreign foods. We don't have a separate system for human food and animal food, and, instead, we have this single system."[19]

Food safety practices in China have come under intense pressure in recent years, with unsafe food continuing to make international headlines. Ninety million Chinese are sickened each year with foodborne illnesses. More than 70 percent of the population lists food safety as its No. 1 concern.[20]

A new amendment to China's Food Safety Law has been proposed to include "stronger regulation of their internal operations and supply chains" and stronger monitoring by government agencies. One analyst believes that this could "bring a rigor to China's food regulatory system that has traditionally been associated with regulation of drugs and medical devices."[21] In the last three years, 2,000 people have been prosecuted in China for crimes related to food safety. Under the proposed amendment, food producers and operators will shoulder more responsibility, and both fines and liability will be expanded. Hand in hand with safety comes an emphasis on wastage; in March 2014, the General Office of the Communist Party of China Central Committee and the General Office of the State Council released a circular targeting food waste and enhancing efforts to reduce waste and loss across the entire food supply chain, including more supervision and inspection.[22]

In October 2013, India's Supreme Court took the remarkable stand of declaring a constitutional right to unadulterated food, directing the Food and Safety Standards Authority in India to "gear up their resources" and "conduct periodical inspections and monitoring of major fruits and vegetable markets."[23] This emphasis on food safety followed the establishment in 2012 by India's government of the National Centre for Cold Chain Development to work in close collaboration with industry and other stakeholders to promote and develop an integrated cold chain in India for perishable fruits, vegetables and other agricultural commodities to reduce wastages and improve the gains to farmers and consumers.[24]

Improved food safety almost always means reduced food loss and waste.

A DIRECT APPROACH TO FOOD WASTE

It has been the "downstream" end of the global food chain, where food retailers and consumers interact, that has experienced the most direct action on food waste from public policy and private initiatives.

> **IMPROVED FOOD SAFETY ALMOST ALWAYS MEANS REDUCED FOOD LOSS AND WASTE**

In a resolution adopted in January 2012, the European Parliament voted to halve food waste by 2025. Recommendations included the introduction of food education courses; the adoption of dual-date labeling to show when food may be sold until (sell-by date) and when it should be consumed by (use-by date); and public procurement rules that favor catering companies that use local produce and redistribute leftover food. One member of Parliament commented, "It is outrageous that almost 90 million tonnes of perfectly fine food gets wasted each year while an estimated 79 million people in the EU live beneath the poverty line and around 16 million depend on food aid from charitable institutions."[25]

Current policy initiatives of the British government also focus directly on reducing waste, which costs England £12 billion annually.[26] The government encourages businesses to build waste reduction into the design of their new products and services. It also works with the Waste and Resources Action Programme (WRAP) on voluntary agreements to reduce food and packaging waste. WRAP also supports a *Love Food, Hate Waste* consumer campaign that provides advice on such topics as recipes and storage tips. In June 2012, WRAP also set up an agreement with the Hospitality and Food Service Sector, with a target for signatories to reduce their food and packaging waste by 5 percent by the end of 2015.[27]

France has taken an especially aggressive stance on retail food waste. As of July 2016, it will be illegal for supermarkets to throw away any food considered edible. This comes on the heels of a 2013 law that required food sellers to label foods to more closely reflect their true shelf life.[28]

In the U.S., the Environmental Protection Agency's (EPA) Food Recovery Hierarchy[29] prioritizes actions that organizations can take to prevent and divert wasted food. Recommendations include reducing the overpurchasing of food, cutting waste in preparation and using proper cooking techniques. Food service providers are encouraged to ensure proper storage and consider secondary uses for excess food. They are also asked to reduce serving sizes and encourage guests to order only what they can consume. The structure of the EPA's hierarchy is to first feed people and then animals. After that, the agency suggests seeking industrial uses for food, followed by composting and anaerobic digestion. The ultimate question posed to retailers and consumers is: How can we divert food from landfills?

A number of federal laws are designed to encourage food donation in the U.S. The nation's tax code provides enhanced deductions to encourage donations of fit and wholesome food to qualified nonprofit organizations serving the poor and needy. The U.S. Federal Food Donation Act of 2008 specifies procurement contract language encouraging federal agencies and their contractors to donate excess food to eligible nonprofit organizations to feed food-insecure people in the U.S. And the federal Bill Emerson Good Samaritan Food Donation Act encourages donation of food and grocery products to nonprofit organizations by providing protection from civil and criminal liability, should the product donated in good faith later cause harm to the recipient. It also standardizes donor liability exposure across all 50 states.[30]

A SEAT AT THE POLICY TABLE

One potentially game-changing public policy initiative is the Green Climate Fund, which aims to support global efforts toward attaining the goals set by the international

ONE POTENTIALLY GAME-CHANGING PUBLIC POLICY INITIATIVE IS THE GREEN CLIMATE FUND

community to combat climate change. Still being organized and funded, it could reach $100 billion by 2020. The fund is targeted at providing support to developing countries to limit or reduce their GHG emissions and to adapt to the impacts of climate change. Its impact on reducing food waste could be significant, if food waste is recognized as a climate mitigation strategy. "The fund is being organized as one of the mechanisms to be used in the climate policy process to help facilitate technology transition and capacity building, especially in growing economies," says Kevin Fay, executive director of the Global Food Cold Chain Council. "We think it's important for this fund to recognize that food cold chain investments are an important environmental investment that have very significant co-benefits."[31]

This is the first time the entire farm-to-fork climate impact of food waste could potentially be addressed on a global basis. What Fay has in mind is leveraging the best practices from around the world, but then tailoring them to meet unique business, cultural and infrastructure needs.

"One big step is to help countries understand what they can do through some simple policies—such as the adoption of dairy standards or shipping standards—that would help provide a basis for cold chain technologies. We also want to help create a dialogue between the industry and these developing countries; a lot of the equipment that has been developed has been developed for different kinds of economies," Fay says. "We need to know on the industrial side the types of technology adaptations that are going to specifically benefit and help facilitate the use of the cold chain in these growing economies."

BREAKTHROUGH FIRST STEP

At the May 2015 meeting of the Climate and Clean Air Coalition's High Level Assembly in Geneva, delegates passed a strategic plan that included a focus on responsible refrigerant management practices and the greening of the food cold chain. As a breakthrough first step, it also emphasized reducing food waste,[32] a sign—along with the promise of the Green Climate Fund—that this important issue might finally have earned a seat at the table of global public policy. If so, it will have quite a story to tell.

For example, we have seen countless ways in which food waste is generated in both developing and developed countries. We have compared its size to a mountain and its climate impact to the largest countries on Earth. We have told the story of food waste around calories, nutrition, biodiversity, GDP, conflict, urbanization and safety. We have established its intimate connection with poverty. We understand its devastating contribution to hunger and malnutrition, and its destruction of our land and water resources. We see that food waste threatens everything from the food security of the poorest smallholder farmers, to the national security of the world's most powerful countries.

In fact, reducing food waste appears to be the only climate policy that lowers GHG emissions, saves water, combats hunger, lifts economies and solidifies national security. Its time has come to sit alongside energy efficiency, renewable energy and more traditional policies as a real, viable opportunity to combat climate change.

Maybe we've said enough. Maybe it's now time to let food waste speak for itself.

[1]Lauren Chambliss, interview with the authors, May 12, 2015

[2, 4, 6, 9, 10, 11, 12, 15]*2014-2015 Global Food Policy Report,* Washington, D.C.: International Food Policy Research Institute, 2015, http://www.ifpri.org/sites/default/files/publications/gfpr20142015.pdf, 2, 6, 41-42, 44, 43.

[3]"The First 100 Days Under Federal Development Minister Gerd Müller," Federal Ministry for Economic Cooperation and Development, 2010-2015, https://www.bmz.de/en/press/The_first_100_days/index.html.

[5]"Second International Conference on Nutrition," United Nations Food and Agriculture Organization, October 2014, http://www.fao.org/3/a-mm215e.pdf, 3.

[7]James Andrews, "The 10 Worst U.S. Foodborne Illness Outbreaks of 2014," December 23, 2014, http://www.foodsafetynews.com/2014/12/the-10-most-harmful-us-foodborne-illness-outbreaks-of-2014/#.VVSxhPlVhBc.

[8]"Foodborne Diseases," World Health Organization, accessed 2015, http://www.who.int/foodsafety/areas_work/foodborne-diseases/en/.

[16, 22]"Food Losses and Waste in the Context of Sustainable Food Systems," a report by the High Level Panel of Experts on Food Security and Nutrition of the Committee on World Food Security, Rome, 2014, http://www.fao.org/3/a-i3901e.pdf.

[17, 18]Regulation No 852/2004 of the European Parliament and of the Council on the Hygiene of Foodstuffs, April 29, 2004, https://www.fsai.ie/uploadedFiles/Consol_Reg852_2004.pdf.

[19]Transcript, March 20, 2014, "FSMA Proposed Rule on Sanitary Transportation of Human and Animal Food, Public Meeting," http://www.fda.gov/downloads/Food/GuidanceRegulation/FSMA/UCM395355.pdf.

[20]"Global Food Safety Conference 2015: Executive Summary, Consumer Goods Forum," March 3-5, 2015, http://www.mygfsi.com/files/Executive_Summary/GFSC_2015_Executive_Summary.pdf, 13.

[21]John Balzano, "Three Things to Look For in Chinese Food Regulation," Forbes, February 5, 2014, http://www.forbes.com/sites/johnbalzano/2014/02/05/three-things-to-watch-for-in-chinese-food-safety-regulation-in-2014/.

[23]Lydia Zuraw, "India's Supreme Court Declares Constitutional Right to Unadulterated Food," *Food Safety News,* October 29, 2013, http://www.foodsafetynews.com/2013/10/indias-supreme-court-declares-constitutional-right-to-unadulterated-food/.

[24]National Centre for Cold-Chain Development, http://www.nccd.gov.in/.

[25]"European Parliament Aims to Resolve Food Waste," *Waste Management World,* accessed 2015, http://www.waste-management-world.com/articles/print/volume-13/issue-1/regulars/news/european-parliament-aims-to-resolve-food-waste.html.

[26]"2010 to 2015 government policy: waste and recycling," Department of Environment, Food, and Rural Affairs, May 7, 2015, https://www.gov.uk/government/publications/2010-to-2015-government-policy-waste-and-recycling/2010-to-2015-government-policy-waste-and-recycling#appendix-4-food-waste.

[27]"Waste Prevention Programme for England: 'One Year On' Newsletter," Department for Environment, Food, and Rural Affairs, December 2014, https://www.gov.uk/government/uploads/system/uploads/attachment_data/file/385049/wppe-1yearon-newsletter201412.pdf.

[28]Roberto A. Ferdman, "France is Making It Illegal for Supermarkets to Throw Away Edible Food," *The Washington Post,* May 22, 2015, http://www.washingtonpost.com/blogs/wonkblog/wp/2015/05/22/france-is-making-it-illegal-for-supermarkets-to-throw-away-edible-food/.

[29]"The Food Recovery Hierarchy," Environmental Protection Agency, accessed 2015, http://www.epa.gov/foodrecovery/.

[30]"Protecting Our Food Partners," Feeding America, accessed 2015, http://www.feedingamerica.org/ways-to-give/give-food/become-a-product-partner/protecting-our-food-partners.html.

[31]Kevin J. Fay, interview with the authors, May 15, 2015.

[32]"Governments, International Organisations, and NGOs Move to Protect Lives and the Climate From Dangerous Air Pollution, May 22, 2015, Climate & Clean Air Coalition, http://www.unep.org/ccac/Media/PressReleases/CCACMovestoProtectLivesandtheClimate/tabid/1060207/Default.aspx.

CHAPTER 14: IF FOOD WASTE COULD SPEAK

To the United Nations:

My name is Food Waste. I noticed that countries are submitting national climate action plans in hopes of reaching an international climate treaty—and I would like to submit a plan of my own.

While I'm not a member of the U.N., if I were a country, my annual 3.3 billion metric tons of embodied CO_2 would make me the third largest in the world, behind China and the U.S. in GHG emissions. My size is also impressive: One-third or more of all food produced is lost and wasted, equaling 1.3 billion metric tons. Think of 1.3 billion healthy elephants happily standing on each other in one pile. That's me.

Given these issues, it seemed important that I submit my own climate action plan, one that will cut food waste to reduce climate emissions. It will also save water, feed more people and boost national economies.

Here is what I suggest.

First, we must start by addressing production and distribution because this is where nearly two-thirds of all food waste occurs. Fruits, vegetables, dairy, meat and fish represent more than half of all the food that is wasted before it can be consumed. All of these foods need basic refrigeration, yet only 10 percent of perishable foods are refrigerated worldwide—despite the fact that refrigeration is the best technology to ensure food safety and security. Technology providers must develop energy-efficient and affordable solutions that farmers and distributors can access, saving crops during transport. If developing countries had the same level of refrigeration for the transportation and storage of food as developed countries, approximately one-quarter of food loss would be avoided.

More food saved also means greater opportunity to use food as an economic resource to bolster national economies. For example, India produces 28 percent of the world's bananas yet represents just 0.3 percent of all internationally traded bananas. With an improved cold chain, the number of bananas exported could grow from 4,000 to 190,000 containers, providing an additional 95,000 jobs and benefiting as many as 34,600 smallholder farms. That would make quite an impact on India's farmers and economy.

Second, we need improved basic food safety standards. This approach would ensure more dairy products and other food would be made safe for consumption. Higher standards would also extend food supplies and prevent waste by encouraging more food preservation techniques.

Third, we should focus on consumers. In developed countries they tend to buy too much and throw it away, or are served too much and can't finish the portions. Changing this behavior can be done through education and awareness. For example, most people might think twice if they understood that each wasted head of broccoli flushes away 5.5 gallons of embedded water. In fact, water is so scarce and valuable it's become the new oil. This is just the tip of the iceberg: By 2025 two-thirds of the world's population will live near a water-distressed area.

Remember me, that 1.3 billion-metric-ton mountain of food waste? If we total all of the water used to grow all this wasted food, it's more than any single nation uses in an entire year.

In addition to lowering climate emissions and saving water, avoiding food waste will help feed more than 800 million people who go hungry every day, which is the equivalent of the U.S. and European Union populations combined. And avoiding the loss of highly nutritious perishable foods, like fruits and vegetables that require refrigeration, will help more than 2 billion people around the world who suffer from nutrient deficiency. Reducing food loss can be an essential strategy to nourish the 2.5 billion people we'll add to the Earth by 2050.

Without a different approach, we'll have to grow more food to feed our growing population. Which means we'll also be throwing more away. A lot more. I'm not sure we have the environmental license and land to do that.

Instead, we should implement readily available strategies to avoid food loss.

So far, traditional climate discussions at the U.N. have focused on energy and power plants. But just as energy conservation can work effectively and economically, so too can food conservation. If funding is made available to developing countries to lower climate emissions, the avoidance of food waste must be on the list. Reducing food waste is the only climate action that unlocks solutions for hunger, nutrition, water scarcity, economic expansion and national security. No other can do the same.

The U.N.'s own FAO quantified the environmental

footprint of food waste and loss in a landmark report released in 2013. All the data needed to act on this issue is available. We must waste less to feed more.

I may not be a country, or enjoy my own standing in U.N. climate meetings, but Food Waste's voice must be heard. The low-hanging fruit for climate protection is literally rotting. Plans like mine to reduce food waste need to be an essential outcome of any international climate accord.

I look forward to working with you.

Sincerely,
Food Waste

ABOUT THE AUTHORS

JOHN M. MANDYCK

John serves as chief sustainability officer for United Technologies Building & Industrial Systems. He chairs the Corporate Advisory Board of the World Green Building Council, and serves as chairman of the board of directors for the Urban Green Council in New York City. John is a member of the Corporate Council at the Harvard University Center for Health and the Global Environment. Follow @JohnMandyck on Twitter.

ERIC B. SCHULTZ

Eric is the former chairman and CEO of Sensitech, an Inc. 500 company specializing in global cold chain monitoring and visibility, and now part of United Technologies. Eric serves as a venture partner with Ascent Venture Partners in Boston. He is author of "Weathermakers to the World," the story of Willis Carrier and the rise of modern air conditioning. Follow @ericebs on Twitter.

ACKNOWLEDGMENTS

In the course of our research we were able to meet with a number of dedicated people who have been working diligently in a variety of ways to help feed the world. Many of these individuals already recognize the issue of food waste as a confounding—but addressable—problem.

Catherine Bertini, former executive director of the World Food Programme of the United Nations and now at Syracuse University's Maxwell School of Citizenship and Public Affairs, shared important insights with us about the relationship between hunger and poverty, and the important role of educating and empowering girls around the world. We're grateful to Marilyn Higgins from Syracuse University for making the introduction.

Dr. Francis J. DiSalvo, Lauren Chambliss, Dr. David A. Dieterich and Dr. Alexander J. Travis of Cornell University's David R. Atkinson Center for a Sustainable Future, and Dr. Julie Stafford of the Cornell Institute for Food Systems led us through a rich and enlightening discussion about food waste, the power of collaboration and a sustainable future.

Kevin Fay is executive director of the Alliance for Responsible Atmospheric Policy, and executive director of the new Global Food Cold Chain Council. He shared important perspectives on greening the cold chain, the role for public policy and how the modern cold chain can reduce food waste.

Rick Fedrizzi, CEO of the U.S. Green Building Council and CEO of the Green Building Certification Institute, was gracious in detailing the work of his organizations in establishing the LEED green building rating system. As a way to protect the planet and drive improved economic returns, green buildings present a compelling model for how addressing food waste through a green cold chain might evolve. Rick also provided early inspiration to write this book.

Brian Greene, president and CEO of the Houston Food Bank, is an experienced leader in the food bank industry whose insights on feeding the hungry in a prosperous urban environment emphasized the fact that food waste and hunger affect rich and poor nations alike.

Gaurav Jain, managing director of Coldex Logistics, helped us to understand the role the cold chain can play in reducing food waste in India.

Charlie Sweat is the former CEO of California's Earthbound Farm, the largest producer of organic produce and specialty salads in the U.S. He was helpful in explaining the options available to California farmers as they cope with the fourth year of that state's ongoing drought.

Jason Turek and Frank Turek Jr. of Turek Farms in New York state were generous with their time, giving us a view of the cold chain from the ground up, starting right from harvest. Jason reminded us that "a lot of people can make a decent crop, but getting it to market safely can be a real challenge." Dan Fessenden was kind to make the introduction to Turek Farms and share insights from his experience in family farming.

Dr. Charles Winkworth-Smith at the University of Nottingham helped guide us through the important work he and his team produced on food waste and nutrition, making clear that it's not just calories but the kinds of calories that count.

Jon Shaw, Sustainability & Marketing Communications, Carrier Transicold & Refrigeration Systems, saw the higher connection between cold chain practices and feeding a growing population before most. He is instrumental in designing research and global events to broaden the dialogue on the future of food.

Finally, we admire and are grateful to all those who dedicate themselves to growing and producing the food that we eat, and we thank those who—in big ways and small—are doing their part to reduce food waste.

We can make a difference.

The Authors